THE FOUNDATIONS OF MUSIC

THE
FOUNDATIONS OF MUSIC

BY

HENRY J. WATT, D.Phil.

Author of *The Psychology of Sound*

Lecturer on Psychology in the University of Glasgow and to the
Glasgow Provincial Committee for the Training of Teachers.
Sometime Lecturer on Psychology in the University of Liverpool.

CAMBRIDGE
AT THE UNIVERSITY PRESS
1919

CAMBRIDGE
UNIVERSITY PRESS

University Printing House, Cambridge CB2 8BS, United Kingdom

Published in the United States of America by Cambridge University Press, New York

Cambridge University Press is part of the University of Cambridge.

It furthers the University's mission by disseminating knowledge in the pursuit of education, learning and research at the highest international levels of excellence.

www.cambridge.org
Information on this title: www.cambridge.org/9781107634411

© Cambridge University Press 1919

First published 1919
First paperback edition 2014

A catalogue record for this publication is available from the British Library

ISBN 978-1-107-63441-1 Paperback

THESE WORKS
ON SOUND AND ON MUSIC
I DEDICATE
TO MY WIFE AND HER ART

PREFACE

In my previous volume *The Psychology of Sound* I made a minutely critical analysis of the elementary phenomena of sound and their simpler complexities, and I developed what seemed to me to be the only systematically true and promising theory of these phenomena. The work was necessarily addressed to those who are primarily interested in such a study, i.e. to psychologists and to physiologists. But I endeavoured to make the material as interesting to the theoretical musician as was possible under the circumstances.

Not that the latter has little interest in such fundamental analysis. On the contrary he is profoundly concerned to know how his art springs from its roots in mere sound and to see that the foundations ascribed to it are such as will evidently suffice to bear the whole superstructure of music. But the purely psychological or 'phenomenal' point of view could not but be new and strange to his mind, requiring some time to come into growth and fruition there. Once the essential nature of the position has been grasped, its spontaneous development is certain.

There is no inherent difficulty in ascribing volume and order to sounds or to tones. The difficulty springs merely from the unfamiliarity of the object in such connexions. At the present day conviction is much more easily secured for descriptions and theories of material objects—even although many of their students may never have come into contact with them at all—than it is for descriptions and theories of psychical objects, although their students are almost of necessity constantly face to face with them at any desired moment. Every one who takes any interest in music has had unlimited opportunities of turning his observations upon tones and their sequences and combinations. But in the great majority of cases he has seldom, if ever, looked studiously at pictures or models of the sensory organ of hearing and in all probability knows nothing of that organ by direct observation of it. But he will nevertheless drink in a description and theory of the material organ with avidity, while he will turn a bored and sceptical ear to a direct analysis and theory of tones, although for both purposes similar methods and explanations may have been used. The mere postulation of a material thing as the bearer of volumes and orders and their coincidences and overlappings seems to bring a special comfort to the mind.

It would be wrong to suggest that this scepticism is quite general in its scope. The artist is certainly clearly aware that what he usually judges and accepts or rejects is the direct phenomenal impression that is immediately before his mind's gaze. But when it comes to science and to theory, what he has learnt to crave for is in the main a material-istic exposition. The musician has long since accustomed himself to a theoretical diet of beats, partials, and material-mathematical expressions for intervals. Such things seem real and tangible, as it were. But, after all, they seem so in most cases only because they are more familiar.

Much of the difficulty is due also to a widespread shallow attitude towards any scientific aesthetics,—an attitude unfortunately greatly encouraged amongst musical theorists by Helmholtz's very unsatis-factory distinction between natural law and aesthetic principles. The mere existence and operation of personally subjective forces that affect our artistic judgments seem to convince so many people that no science of these judgments can be achieved. There is no disputing about tastes, they say. And if a body of critical knowledge can be extracted from established works of fine art, such rules are held to be merely the conventions of the ages that created them. The next genius that comes along may blow the whole system to the winds of oblivion. So it seems to those who are struck most of all by the innovations of each master and have not perhaps the patience to follow out the great purpose that is common to them all and that each merely carries on to finer and finer issues. But a great master is by no means an accident. He is one who, taking himself as a man amongst many, has learned how to construct an enduring object that is far more likely to arouse in others the joys of beauty he has felt and anticipated for them than are the works of lesser minds. His medium is the orderly realm of mind that he shares with his fellows. But its laws are not his creation; they are only his discovery. He has learnt to turn them to his will. The scientist who comes after him has, with his help, to formulate them in knowledge. That knowledge is the science of aesthetics.

In this volume I have sought more or less evenly to serve the purposes of both the psychologist and the musician. In order to make the work complete in itself up to a certain point I have traversed the ground covered in the psychological part of the earlier volume, omitting only those parts that are of little interest to the musician. All critical dis-cussion has been passed over at this stage, so that the earlier part of this volume is more or less a careful and straightforward ('dogmatic') exposition of the fundamental notions of the psychology of tone. I think the musician should find it useful and helpful, as also will those who wish to have an exposition of the system without the technical discussions and criticisms. Those who are familiar with the previous volume will hardly find anything new before chapter IX (p. 55). Except in so far as they are interested in straightforward and logical exposition, they may begin the work at that point. The previous pages, however, are not in any sense a mere repetition of the earlier volume. They have been written entirely afresh. Only, as is after all inevitable, they are based on the same body of facts and notions as was the analytic psychological work of the earlier volume. I have not gone into any binaural or physiological problems this time.

The parts beyond chapter IX are addressed both to psychologists and to musicians. In the preface to the previous volume I said that my "theoretical constructions must be carried somewhat farther before they can be held to have passed fully over into the elements consciously used [and known] by productive musicians and appreciative listeners.... The working musician definitely takes over at a certain point the raw materials of his art from the real psychical processes of hearing, inaccessible in full to observation, and then proceeds to construct from them vast new realms without consulting anything that lies beyond the ken of observation." In this volume I think I have succeeded in carrying the psychological groundwork of the previous one forward so as to bridge the gulf between the psychological elements and processes of music on the one hand and on the other the sensory stuff and functions of music as the musician observes them. If my results and analysis are valid, the musician should now have a nearly complete and sure basis to work upon, that will give a scientific foundation to all his elementary observations and satisfy him with a sense of firm ground upon which to build.

The work of carrying the psychological analysis thus far has not been light. For as things have been till now, the probability of any one person being equally and fully conversant with the science of psychology and with musical history and theory was exceedingly small. I am aware of the great deficiencies in my own preparedness for the latter half of this great double task. I have tried to make up for want of experience by keeping closely in touch with the general trend of the judgments of ripened masters in musical theory of the empirical analytical order. I feel sure that results have offered themselves to my hand that theirs passed by, however closely. And I am certain that they will not be loth to recognise the validity and usefulness of these results. I only wish that they would feel impelled to make themselves familiar enough with the results of psychological work on sound to carry the new and promising basis of theory far forwards into their own fields. In order to facilitate this junction and continuity of work I have ventured as far out into musical regions as the material accessible to me will allow. Analytic musicians will certainly be able to carry things much farther on until the new outlook permeates the whole theory of their art.

There is still much to be done. Although I feel a growing assurance that the lines of analysis I have followed lead in the right direction, I still feel, not only that the elements of analysis are not nearly well enough assimilated to one another, but also that a much more detailed and exact foundation is required for their final acceptance than the empirical generalisations of writers on harmony afford, however correct and worthy these may be. A better form of evidence is required; and the best I can imagine would be a statistical study of the great composers, seconded by systematic experiments with the best observers.

In order to bring my results the more quickly into touch with practical issues and exposition, I have again ventured upon an exposition of results in the most familiar terms of elementary musical knowledge. The underlying idea of this account is the construction of a simple framework round which an introductory account of the basis and rules of harmony might be built.

Some text-books still maintain the effort to link a system of harmony to the traditional lines of scientific explanation founded upon the harmonic partial tones of musical sounds. Of this E. Prout's splendid treatise is in its earlier form a most notable example. But the effort has been a complete failure, whose only result must have been to bewilder any student of a logical turn of mind. Many other writers have abandoned every sort of scientific introduction; and Prout has followed them in the later edition of his work. That course seems, on the whole, preferable to the former. But the bald exposition of analytic generalisations, however true to the great masters they may be, can never be enough. The mind craves for some logical nexus to give the whole mass spontaneous life. Even analytic exposition can never be complete until it has developed into a logical system whose fundaments bear the higher refinements as a tree bears its fruit. And these refinements can hardly be properly approached and stated until the system inherent in the products of analysis has been discovered.

I hope the lines of introduction I suggest will help to relieve the beginning students whose musical mind is not good enough to be a test for every effect and a storehouse for every impression, of the perplexity and confusion that so soon overwhelms his first efforts at 'harmony.' And perhaps the teacher will find comfort and stimulus in having (I hope) a good explanation to offer, by which the student's logical mind may be brought to the support of his natural gifts of ear.

It may seem strange to suggest that the exposition of the theory of harmony has hitherto been devoid of system. There has certainly been no lack of desire for system and of effort to form it. Dr Shirlaw has recently given us a lengthy account of "the chief systems of harmony from Rameau to the present day." These are very numerous and of the greatest diversity. But it is evident from a study of them that they are all mere castles in the air, as it were. They lack, one and all, any proper sort of foundation. Each man's construction is like a toy castle built upon his outstretched hand. The parts are in the main merely laid down beside and upon one another. The weight of the upper parts repeatedly pulls out the joints from their places; for they are not bound by any mortar. And the builder's own care and anxiety does the most to precipitate the fall of the pile. Perhaps in the end he throws the whole game into the fireplace and warms himself by it while he reflects on what useful thing he might do instead.

Dr Shirlaw, it is true, in spite of the extensive criticism he bestows upon these theorists, still believes that a good foundation for musical theory is to be found in the "resonance of the sonorous body." "As the sounds of this harmony are contained in the resonance of musical sound itself, all harmony has its source in a single musical sound" (60, 481). But the day is well past in which a system can hope for a moment's success that rests upon such philosophical naivety. The scepticism of Prout and others is far more worthy and hopeful.

The sonorous body is out of the question. Between it and music there lie not only the phenomenal material of sense but the mechanisms of the ear as well. The ear must stand in real systematic continuity with the physical processes of sound, including as a mere part the musical sonorous body; the phenomena of sense must bear a similar relation to the ear. The problems of musical science consist primarily in showing the system of bonds and relations by which the stuff of auditory sense is built up into music that delights the soul. There are doubtless various other ways of building up structures of sound than the musical one. But the latter is one of the most important. In it the aesthetic test is applied; a process of selection is set over nature's indiscriminating generosity. But it is obvious that the aesthetic test no more accounts for the auditory material of music than a five shilling entrance fee accounts for the audience that appears at a concert.

As far as the sensory stuff of music is concerned, then, and apart from such things as rhythm and musical form, no tenable theory either of the auditory basis or of the aesthetics of it has ever yet been advanced. But there does exist a considerable body of generalisations that have been won from the works of the masters by analytical induction. By this I do not mean merely the useful attempts that have been made to form systems of chords; for these have been largely vitiated by the attempt to derive the chords from some fundamental chord or other, when it was evident for strict logical thought and intuition (if one may say so), that all the time no satisfying criterion existed by which any such fundamental chord could be established. D'Alembert expressed this thought in relation to Rameau's efforts when he said they yielded no demonstration, but only a system. I mean rather the body of experience that has been known as the rules of part-writing and of the formation of melody and the factors that modify the operation of these rules. These rules have been the step-children of musical theory, ignored and despised as far as any system building was concerned, treated only as accidents of the all-absorbing science of chords and their origins. But this despised child of musical empiry may well take precedence in musical science and end by being the queen who will show her sisters in music what their functions really are.

I am far from suggesting that all the main problems have been solved. But it is good that lines begin to appear that seem promising and strong. Many must labour at the task before it will ever be wrought to the satisfaction of all. Still, there seems to be no reason now why rapid progress should not be made, even if no one would be quite so hopeful as Kant was about his efforts as to be sorry for the next genera- tion that would have nothing to do but learn the results.

When one surveys the actual changes in the fundamental notions of a subject like music required by even very extensive inductions and analyses, they may seem to be surprisingly small. This is perhaps most striking in the problems of consonance and dissonance. It is not so much a striking change of substance that is produced but a far-reaching improvement in the body-building capacities of the fundamental nucleus. As cytology shows us, trifling changes in the numbers and arrangement of chromosomes may alter the resulting organism profoundly. And ages are required to attain these trifling changes. So in science : some fundamental notions are lethal, they can compound to no living organism; others very slightly different grow and reproduce with great vigour. That I hope may be true of my own analyses. And as great interest attaches to the way in which new notions of promise come to light, I have not sought to obliterate the traces of this process in my writing. A straightforward factual and logical development naturally expunges all these things. But that method really takes much for granted and is only to be used when men are already well disposed to be convinced by the established ways of a science.

I am indebted to Prof. W. B. Stevenson for a summary of de Pearsall's pamphlet in the British Museum, and to my brother, Rev. T. M. Watt, for reading the proofs.

H. J. W.

30th March, 1918.

CONTENTS

		PAGE
	PREFACE	vii
CHAP.		
I.	The reduction of instrumental tones to a single series of pure tones	1
II.	Analytic description and theory of the series of pure tones . .	5
III.	Degrees and theory of consonance and dissonance (fusion) . .	15
IV.	The relations of fusion to beats, partials, and difference-tones . .	23
V.	The consonance of successive tones	31
VI.	The nature of interval	36
VII.	The musical range of pitch	45
VIII.	Our point of view towards the auditory field	48
IX.	The relative importance of synthesis and of analysis . . .	55
X.	The equivalence of octaves	65
XI.	Consecutive fifths	81
XII.	The system of facts regarding consecutives	98
XIII.	The reason for the prohibition of consecutives	109
XIV.	Exceptions to the prohibition of consecutives	115
XV.	Hidden octaves and fifths, etc.	122
XVI.	A fourth from the bass	133
XVII.	Common chords or concordance	141
XVIII.	Melodic motion in relation to degrees of consonance . . .	151
XIX.	Melody (or paraphony) as the primary basis of music . . .	160
XX.	The factors that modify paraphony	169
XXI.	Retrospect and the outlook for theory	185
XXII.	Synopsis or outlines of instruction	198
XXIII.	The objectivity of beauty	214
XXIV.	Aesthetics as a pure science	221
	WORKS CITED	233
	INDEX OF AUTHORS	236
	INDEX OF SUBJECTS	238

We shall now proceed to the consideration of Harmonic and its parts. It is to be observed that in general the subject of our study is the question: In melody of every kind what are the natural laws according to which the voice in ascending or descending places the intervals? For we hold that the voice follows a natural law in its motion, and does not place the intervals at random. And of our answers we endeavour to supply proofs that will be in agreement with the phenomena—in this unlike our predecessors. For some of these introduced extraneous reasoning, and, rejecting the senses as inaccurate, fabricated rational principles, asserting that height and depth of pitch consist in certain numerical ratios and relative rates of vibration—a theory utterly extraneous to the subject and quite at variance with the phenomena: while others, dispensing with reason and demonstration, confined themselves to isolated dogmatic statements, not being successful either in their enumeration of the mere phenomena. It is our endeavour that the principles which we assume shall without exception be evident to those who understand music, and that we shall advance to our conclusions by strict demonstration.

ARISTOXENUS (cf. 1, 188f.).

Symphonic are those in which, when they are simultaneously struck or blown on the flute, the *melos* of the lower in relation to the higher or conversely is always the same or (in which), as it were, a fusion in the performance of two tones occurs and a kind of unity results.

Diaphonic are those in which, when they are simultaneously struck or blown, nothing of the *melos* of the lower in relation to the higher or conversely appears to be the same or which show no sort of fusion in relation to one another.

Paraphonic are those that, standing in the middle between the symphonic and the diaphonic, yet appear symphonic when played [on instruments, or "in heterophonic passages on instruments" (11, 139)]

GAUDENTIUS (cf. 54, 69).

CHAPTER I

THE REDUCTION OF INSTRUMENTAL TONES TO A SINGLE
SERIES OF PURE TONES

THE world of sound is bounded by the two extremes of pure tone and mere noise. The home of music lies in the lands around the ideal of tone. This ideal forms the first problem of musical science.

It is properly termed an ideal because pure tones rarely, if ever, occur under natural circumstances. That is evident from the familiar fact that, however perfectly a musical instrument may be played, its tones are easily distinguishable from those of other instruments, even though they may be of the same pitch. The same series of tones of exactly the same pitches, e.g. the diatonic scale on a c^1 of 264 vibrations per second, may be given by an indefinite number of instruments, and will be recognised as different on each, in spite of the sameness that is obviously common to all.

This peculiar complexity of tone has been explained by modern research in a way that at least in principle is complete and final. However pure and beautiful an instrumental tone may be, it can be analysed into audible parts by special means of two kinds. In the first the ear is provided with instruments which will increase the intensity of certain parts of the tone—if they are present in the tone to be analysed—beyond that of the other parts. When the resonator is placed against the ear, it seems to be full of the magnified sound, whose pitch may be surprisingly different from that of the tone it comes from. But its presence in the resonator is easily shown to depend upon the studied tone. Whenever the resonator is placed against the ear, it appears; and it only appears in the resonator when the tone being analysed is sounded, unless, of course, it is given from some other source at the same time. Any such experimental error can be easily avoided in most cases. After a little practice with the resonator, the partial tone will often be distinguishable in the whole tone even when the resonator has not been placed upon the ear. This is not the result of imagination or illusion. It only means that the ear has now been trained for this particular case to expect a certain partial tone and to direct observation specially upon it, so that it appears to be more or less abstracted from the whole.

A generalisation of this procedure gives the second method of analysis. The ear is first prepared for special observation (or abstraction) by listening for some time to the tone expected to occur. The tone to be analysed is then presented and if it contains the prepared tone as a partial, the latter will probably be heard sounding faintly through the whole. If the ear has been prepared for a tone whose pitch is not quite the same as that of the partial actually present in the whole, the listener will not hear the pitch he expects, but another that lies in pitch near the one expected. When the ear has thus been trained for many or for all the partials of a tone, it may be able to run through them all in sequence without any special preparation. And in the course of time it may learn to do this for any sound of a tonal nature. Even then, however, the tones of well played musical instruments do not cease to be the beautifully perfect unities they were before. They do not fall to pieces, as it were, permanently, but only when the attention is concentrated and moved from one of the partials to another.

The occurrence of these partials is due to the fact that most musical instruments when brought into a certain rate of vibration—n times per second,—fall at the same time into various rates of vibrations that may be any whole multiple of $n : 2n, 3n, 4n, 5n$, etc. The pitches corresponding to these ratios of vibration are : octave, octave and fifth, double octave, double octave and third, double octave and fifth ($6n$), a pitch slightly flatter than the $b\flat$ above ($7n$), the triple octave ($8n$), then the d ($9n$) and the e ($10n$) above this c, and so on. The pitch of any partial may easily be reckoned out from a knowledge of the ratios for the chromatic scale and by approximations thereto. These ratios are :

c,	$d\flat$,	d,	$e\flat$,	e,	f,	$f\sharp$,	g,	$a\flat$,	a,	$b\flat$,	b,	c^1
24,		27,		30,	32,		36,		40,		45,	48
†	$\frac{16}{15}$	$\frac{9}{8}$	$\frac{6}{5}$	$\frac{5}{4}$	$\frac{4}{3}$	$\frac{45}{32}$	$\frac{3}{2}$	$\frac{8}{5}$	$\frac{5}{3}$	$\frac{16}{9}$	$\frac{15}{8}$	$\frac{2}{1}$

All of these ratios can be derived from those of the octave ($2 : 1$), fifth ($3 : 2$), and major third ($5 : 4$) that have played so important a part in the history of musical theory.

The existence of partial tones has been confirmed objectively in a number of ways. In various cases the presence of vibrations corresponding to the pitch of the partials heard can be demonstrated to vision. A long stretched string may be seen to vibrate, not only in its whole length, but also in parts of such length as would give the various partials as independent tones, if these parts of the string were made to vibrate separately from the rest of the string (cf. 35, 91f.). In other cases the motion of a minute mirror standing in connexion with a vibrating membrane may be photographed with so little error that the result may be taken as representing the motions of the air that excite the vibrating membrane (40, 78 ff.). (Phonography is dependent upon such a vibrating membrane, and everyone is aware how good a reproduction of sound may be obtained thereby.) When the photographs so obtained are subjected to mechanical 'harmonic analysis,' the partial tones that result correspond very closely with those that can be heard by the most careful analysis with resonators or with prepared attention.

Such studies as these show that tones of the same nominal pitch from different instruments differ only in respect of the group of partials from the full series (of possible multiples of n) that they contain and in the relative strength of these. Some tones like those of tuning-forks and of the flute contain very few partials, perhaps only the first. Others, such as pianoforte tones, are rich in the lower partials. Others again, like those of the trumpets, contain a host of high partials in great strength, which give them their peculiar brightness and brilliancy. And so on. The results of this line of study will be found extensively in special treatises and in text-books of physics (40, 175 ff.; 20, 118f., etc.).

Partials may be eliminated from the tones of any instrument by the method of physical interference. That consists in principle of the conduction of a sound containing at least one partial other than the fundamental component of the tone, along a tube which for a certain length is double and then unites again to enter the ear. The one doubled part is made longer than the other by half a wave-length of the partial to be eliminated. When the parts unite, each will bring this partial in exactly opposite phase to the other. If the one is at the phase of maximal condensation of the air, the other will be at that of maximal rarefaction; and the result will be the elimination of that component of the aerial disturbance. If there are many partials in the tone, there will have to be a special device of this kind for the elimination of each, so that the apparatus for the production of a pure tone from an instrumental one is apt to be somewhat complicated, unless special care is taken to begin with a tone containing very few partials, such as that of a tuning-fork.

In spite of the difficulties that thus face any complete generalisation, no one doubts for a moment that the series of perfectly pure tones, each consisting of a fundamental with no upper partials at all, thus isolated, is one and the same, no matter from what instrument it may have been derived. This is in perfect accordance with the results of the attentional analysis of instrumental tones. For the partials separated by the attention do not seem to differ from the (pure) tones of identical pitch, otherwise produced, in any such way as would make us believe that the series of pure tones is not the same whatever its source may be.

Thus we obtain a simpler starting-point for our study of tone—the series of pure tones, each one of which corresponds to a certain fixed rate of aerial vibrations, unmixed with any other rate. Now this series is perfectly continuous. If we start from any ordinary pitch, e.g. middle c of 264 vibrations per second, we can raise or lower the pitch of tone gradually, producing differences as minute as the mechanical means at our disposal will allow. There is no reason in the nature of tone why we should select any one pitch or rate of vibration for our c or a rather than any other. And even when a standard pitch has been adopted for practical purposes, minor variations due to change of temperature, mis-tuning, etc., are inevitable. It has been claimed that the vibratory rate of 256 should be taken as the standard of 'philosophic pitch,' because $256 = 2^8$; i.e. if we imagine a tone of one vibration per second, the (fictional) tone of two vibrations would be the octave of it, four vibrations would give the double octave, and so on, so that the eighth octave would give us 256 vibrations (50, 33 ff.). It is certainly very useful to have a commonly accepted standard for convenience of reference. Then we know what rate of vibration is implied by any nominal pitch, e.g. A_1, without having to give it separately. But the standard now perhaps most commonly in use is a c^1 of 264 vibrations per second. One advantage of this basis (although a slight one), is that it is a multiple of 24, and so can be readily used in connexion with the diatonic series of ratios stated above. I shall use this standard throughout the following pages unless some other standard is specially indicated. The usually current nomenclature of octaves may be looked upon as starting from 'middle c,' the c common to the baritone and contralto voice, which is called c^1. Above that the octaves are c^2, c^3, c^4, c^5 (the highest note on the large concert grand piano), c^6, etc.; below we have c^0, C, C_1, C_2 (A_2 is the lowest note on the same instrument). Plain letters will thus indicate absolute pitch, italicised letters relative pitch.

CHAPTER II

ANALYTIC DESCRIPTION AND THEORY OF THE SERIES
OF PURE TONES

HAVING thus reduced the usual tonal material of music to its simplest components, we have now to describe this continuous series. The terms of our description, to be scientifically useful and explanatory, must be such as will bring tones into systematic connexion with as many other similar objects as possible.

Being dependent upon the working of a sense organ—the cochlea of the ear,—tones are classified in psychology as sensations. We naturally expect them to show great similarity to the sensations we get from our other sense-organs, such as the eye (vision), the tongue (taste), the skin (touch, temperature, pain), and various others, such as hunger and thirst. The similarity of all these to one another is certainly not at first striking. And it has usually been thought that the differences are far more numerous and important than any resemblances there may happen to be. Many men, indeed, judging by the perennial failure of the attempt to bring our different senses into systematic connexion, have adopted a standpoint of extreme scepticism towards any such claim or expectation. But since psychology became, some decades ago, an experimental science, the study of the sensations has been pursued most carefully and exhaustively, and the real relations of resemblance and of structure between our various sensations have gradually grown clearer. Beyond this nothing is required but a frank and determined rejection of the old prejudice and a whole hearted effort to work out the inner similarity of sound and of the few other senses we have. The problem then is to describe the tonal series so as to show the inner connexion not only between all the parts of the series, but between tones and the sensory objects of the other senses.

The first and most obvious feature of sounds is that which distinguishes them from the sensations of other senses. No kind of sound or group of sounds is ever confused with a sight or with a touch. Psychologists call this a difference of quality. The word quality is often used by musicians to designate that difference of tones of the same pitch which is due to the peculiar blend of partials they contain. It is better, however, to call this the (pitch) blend of tones. For practical purposes that word is the best which most readily suggests the thing named, or its cause, or the like. The word blend is in common use as a name for similar differences in objects that appeal to other senses, especially to taste and smell, of which the latter is the more important. The blend is here due to the mixture of the components. Similarly the blend of a tone will be the difference due to the admixture of partials, which a trained ear can learn to pick out and name, as a practised palate will detect the components of a tea or the varying flavours of a wine. This word 'blend' seems better than the French word 'timbre,' which does not fit into our language either in its native pronunciation or in ours.

The second attribute that is found in all tones and in the sensations of other senses is intensity. Both scientific and popular usage agree as to the meaning of this term. The word loudness is not so useful for classification, because it is inapplicable to the other senses and so does not serve to indicate any variant common to them all. One word of frequent occurrence must be carefully avoided in this connexion, namely, volume. We think of loudness as great volume when many instruments sound together as in an orchestra and so make a very intense sound or a mass of very many sounds. Having thus associated many sounds with much sound, we often use the word where there is obviously only one sound present, as when we speak of the great volume of a singer's voice, especially of a contralto's or a bass's. Here a touch of the scientific usage begins to appear. But that highly justifiable usage does not tolerate any confusion of volume with loudness.

Volume is properly used to distinguish that difference between tones of different pitch that makes the low tone great, massive, all-pervasive, and the high tone small, thin, and light. The other words we use to designate differences of pitch have the same sort of association. Sharp and flat are closely akin to thin and broad, or small and large. The Latin and French words gravis, grave, acutus, aigu, bear the same implications. In the eighth Problem on Music Aristotle asked: "Why does the low tone dominate the higher? Is it because the low is the greater? For it is like the obtuse angle, while the other resembles the acute angle." And the twelfth Problem answers: "Is it because the low tone is great and therefore more powerful and because the small is included in the great?" (65, 17, 19; cf. 16, 13, 19).

Although the words of these sentences are very suggestive to a theorist of the present day, it is doubtful whether Aristotle meant really to ascribe differences of size to the tones as mere sounds or sensations. His mind was very much impressed by the discovery that the low string gives out not only its own tone but the higher octave, so that, as we should say now also, the low tone contains 'the higher one'—its own first higher partial, the octave of itself. But Aristoxenus refers to "the blunder of Lasus and some of the school of Epigonus, who attribute breadth to tones" (1, 167). And many modern writers have inclined more or less tentatively towards this idea as an explicit description of tones as such. We must now certainly take the idea with complete seriousness and think of tones as of different size or mass or bulk, just as a visual sensation can be of different size in respect of its mass or area, or as pain and hunger can be large and massive, or as pain and touch can be small as sand or needles. There is every reason to believe that this difference of volume or extent is dependent upon the number of elementary sense-organs of hearing that are in action at the same time. But that is a question for physiology.

The only other property of pure tones is what we commonly call their pitch. By pitch tones fall into a definite order or series. This is not naturally a discrete series: like the ordinal numbers, of which each one is an individual separated from the next by a unit of space, into which other numbers of a fractional nature may be fitted; or like the pitches of the diatonic scale. It is a continuous series: we can pass from any one point of it to any other by gradations that are not distinguishable from those that lie next to them on either side, but that are distinguishable from those that lie a certain distance away on either side. And the series is ordinal, not because it can be considered *conceptually* as a continuous series of positions, but because it appears so to us phenomenally, as is often said, or merely as sensation. The series of colours of the spectrum, merely as colours (i.e. apart from their position in the dispersed spectrum, and from the wave-length they depend upon) can be treated in an ordinal way, although that series is, as sensation, really a series of qualities. But the pitch series is, as sensation, itself really ordinal[1]. It presents itself to us as ordinal and

[1] Aristoxenus wrote: "Tension is the continuous transition of the voice from a lower position [τόπου] to a higher," etc. (1, 102. 172; 14, 83 f.). The Greek term is highly suggestive. But it does not seem certain that he meant by it more than cessation of change of the voice, a permanence of one kind of activity. It is, of course, significant that in this, case we naturally incline towards the term 'place' or to the idea of the voice's 'moving.' We do not incline to say that the weather 'moves,' or the colours of leaves 'move' in the autumn, when we mean only that they change. No doubt Aristoxenus used the terms 'place' and 'motion' at the suggestion of the ordinal and motional aspects of tone. Nevertheless his concepts of position and motion of the voice probably did not include more than what he might have attributed to a thing that only changes, i.e. progress of change and arrest of change Cf. 80, 221 f.

calls for ordinal names, whether we know anything about the wavelengths that cause it or not.

The attributes of tones thus far enumerated are: quality, intensity, volume and pitch. The relations between these four are an important problem. It has been suggested that intensity is a sort of density of sensation, as it were. Just as a gas may fill a certain volume and yet be very thinly scattered throughout it, so it is thought a sensation may be of one and the same quality, and volume, and pitch, and yet be more or less dense—or intense. Suggestive arguments in favour of this view have been advanced, but they do not yet seem sufficient for their purpose.

The relation between pitch and volume is much clearer. When tones are compared with noises, a marked difference is apparent. Tones, as everyone feels and knows, are smooth and regular, noises are rough and irregular. Tones may also be said to be balanced and symmetrical, while noises are chaotic and disorderly. These descriptions obviously refer to the volume of tones, not to their pitches. A pitch has only a definite position or place; it is not smooth or balanced. But pitch gives tone a position as a whole; it is by means of pitch that tones are brought into a definite and accurate series, and their volumes along with them. The question then arises: what position has pitch itself in the tone's volume?

This is not an absurd question, but a very natural one. For if tones have an aspect of volume and can be arranged in a very definite and single series by means of pitch—a property that is distinguishable from volume; and if pitch is not only thus really ordinal, but is also felt as ordinal or appears to us so as a property of sensation; it is perfectly natural to suppose that what is thus ordinal is a part of the tone's volume and to ask in consequence—which part of the volume constitutes the pitch?

Of course, one's habits of thought may oppose this line of inquiry. One of the greatest obstacles to the advance of knowledge is the opposition our minds offer by the mere force of unfamiliarity to the application of old and simple notions to common objects to which they have not hitherto been applied. The mind seems to refuse to establish the desired connexion. All sorts of excuses and objections are offered to the new invitation. "Metaphors are so misleading." But it is not a case of metaphors now. Pitch is no mere analogy; the ordinal status and arrangement of tones is one of the bed-rock facts of music. And 'volume' is no more a mere simile than is interval or concord or discord. It is as much there as any fact could possibly be. Psychologists admit it more and more frequently, and it is only a matter of time till everyone who considers the subject will agree with them. Nor is it 'mystical' to suggest that pitch has a position in volume. A line of well-founded and logical thought is only mystical to those who do not take the trouble to follow it carefully. A mystic is one who claims to have special insight or experience which he has discovered by accident or providential good-will and which he is powerless to reveal to others either because it defies all description or because, not knowing how he himself attained it, he is unable to lead thither all who would share it with him. But there is nothing mystical about pitch or tonal volume; nor are the ordinary logical processes of inference held to be the special privilege of a few minds.

We may therefore consider our question clear and reasonable. And the most likely answer follows naturally from the apparent balance and symmetry of tones as compared with noises. We may assume that pitch holds a central position in volume. And, as pitch is ordinal, while volume suggests a volume of parts or particles, we may go on to assume that pitch is constituted by a specially prominent or noticeable part of the volume of sound that makes up a tone. We certainly do not hear tone as a group of distinguishable particles like a handful of sand. We hear it as a continuously smooth closed volume. But nevertheless a part of this whole volume might well be more noticeable than the rest, just as a part of a variably 'toned' visual surface may be most deeply coloured—red, for example,—although we could not pick out and isolate any part of it that would be all, and nothing less or more than all, the reddest part. Yet no one doubts that a visual surface consists of a mass of minimal particles of colour surface, grouped into a continuous whole. These particles are presumably the minimal areas of colours given by single recipient visual organs—the cones (and rods) of the retina. There are also in the ear elementary receptors of sound; and these presumably afford us the minimal particles of sound that make up tone.

Now all (pure) tones are the same in symmetry and balance and smoothness. So we may consider this central position and predominance of pitch to be characteristic of pure tone as against all grades of noise, which are relatively rough and unbalanced, vague or indefinite in pitch or marked by many prominent points of pitch. Tones differ from one another in size of volume and in the ordinal position of their pitches relatively to one another. The pitch of a higher tone lies a little to one side of the pitch of a tone just lower in pitch; and the pitches of all tones together form a single linear series, having the tone of greatest volume at one end and the tone of least volume at the other. If we were to project the volumes and the pitches of all the tones of the series against one another in our thought, we should obtain a scheme of the following kind:

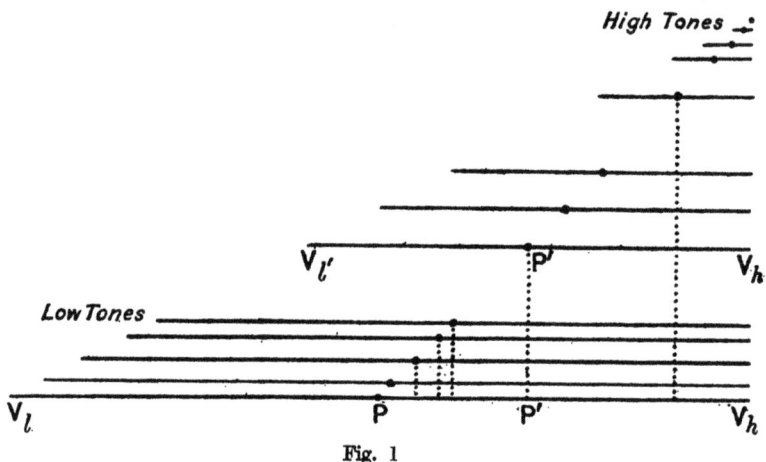

Fig. 1

If we placed all the pitch-points on a perpendicular line above one another, we should indeed represent the decrease of volume (as we go up) properly by the decrease in the breadth of the line used, while the symmetry and balance of tone would be indicated by the central position of the P point in the volume line (Vl = 'lower' end of volume, Vh = 'higher' end of volume). But we should not have given any representation of the fact that the pitch of a tone higher than another lies on one side of the pitch of the latter in an ordinal series. This series is quite properly indicated in our figure.

We have as yet no proof for the assumption that the Vh ends of all the volumes should lie perpendicularly above one another, or— whether in mere projection on the base line of the figure (which may be supposed to represent the greatest possible volume or the lowest possible tone), or in reality—should be the same point. It is conceivable that the pyramid of tones should be acute or obtuse angled rather than right-angled. But these alternatives are far from likely for various reasons of which the most important will be set forth immediately.

It would follow from our scheme that if the projected series of pitches is in any way real, that ought to appear in an unmistakable manner when tones of different pitches are given simultaneously. And this is the case. Simultaneous tones seem to be mixed together or to fuse with one another or to intermingle. They never appear to be entirely apart from one another as two patches of colour often do, when they are separated by a patch of a third colour or when they just bound one another. Each patch of colour is seen as well when both are given together as when they appear singly and successively. Not so two tones; they always appear to be in each other's way, to crowd upon or to overlap one another. This holds even for the greatest extremes, when the highest and lowest tones are given together. And yet at the same time the two tones by no means completely lose their individuality or become indistinguishable in this 'mixture,' as two colours do when they are mixed by being cast upon the same surface or by the rotation of them on a disc. Blending of colours gives a new colour in which the components are essentially indistinguishable by any one who did not see them before their mixture. But musical folks can detect the component tones of a chord with ease and certainty. In spite of the overlapping or interpenetration of tones they are more or less readily distinguishable. Their pitches generally strike us as being the same in mixture as in isolation.

If, as we have supposed, the series of pitches represents a series of real particles of sound, a ready explanation of this peculiar intermingling of tones is at hand. Then the sounds that form the highest musical tone will form the last (or highest) part of the volume of all simultaneous lower tones. Being the same sounds, the two common parts will overlap or intermingle; but, as the higher tone adds its intensity to the high part of the low tone, and especially the predominant intensity of its pitch, the high tone will stand forth, and be detectable in the lower tone in spite of the overlapping of the two. And the rectangular shape of the scheme we have figured is justified. The possibility of an obtuse-angled figure is then excluded; for that would imply that the volume of the highest tone lay beyond the volume of the lowest tone, so that a medium tone would not mix with, or obscure, a very high tone in the slightest degree. And the probability of an acute-angled form being the true scheme is equally small; for in that case the movement of pitch that accompanies the continuous decrease of volume would proceed, as it actually does, steadily in one direction, but only up to a certain point, after which the direction would be reversed. Of this there is no actual trace in hearing. We may therefore for the present safely follow the scheme depicted in the figure. We shall obtain further evidence on this point when we come to the study of the grades of fusion.

We conclude, therefore, that tone is a mass or volume of minute (hypothetical) particles of sound sensation, of which those at its centre are the most intense, while the others grade themselves on either side in the whole volume so that the mass appears regular and balanced. Since in a pure tone no other predominating points or pitches appear, we must suppose that the intensity of its constituent particles decreases gradually and regularly from the central pitch towards the limits of the volume[1]. In this respect every tone is appreciably the same, no matter what its pitch may be. The particles that constitute volume must be called hypothetical, because the volume does not appear to be a group of distinguishable parts. We do not separately experience the minimal particle except possibly in the case of the tone or sound of the smallest possible volume, i.e. the highest audible sound. That may be approximately the minimal particle. Between this highest particle (or pitch) and the pitch of any tone there lies the whole series of pitches that leads from the latter to the former. And we have reason to infer that half of the volume of any tone is made up of the pitch particles that appear in the series of tones progressively higher than itself up to the highest tone. So the lowest audible tone would consist in its one ('upper') half of the whole series of audible pitch-points. The other (or 'lower') half of its volume is not used for the formation of the pitches of other tones. That is, of course, no argument against its existence.

Every tone must, therefore, contain within it the volumes of any higher tone. And all tones are constituted from a single series of sound particles, of which they incorporate a series always beginning at the common upper end and stretching downwards as far as the size of each tone's volume requires. These and other such relations can easily be read from the figure already given.

[1] Wm. Gardiner in his popular compendium of musical topics, *The Music of Nature* (13, 188) gives a curious diagram entitled: "The wind instruments—the shape and order of their tones from the lowest to the highest," which shows a column of twenty-one oval figures, coloured differently for the different instruments, with a small dot at the centre of each. These ovals grow gradually smaller towards the top of the column. They might all be inscribed in an angle whose sides were about five centimetres long and two-thirds of a centimetre apart at the ends. Gardiner did not in any way elucidate in words the reasons that led to this diagram.

The series of pitches may be said to define clearly one-half of *one* of the dimensions of tonal volume. We say 'one,' because we do not feel tone to be a mere line or length, but a volume; something areal or massive or round, as it were. Of course, in so far as we think of tones from the pitch point of view they fall into a perfectly definite ordinal or linear series of no thickness at all, so to speak. Yet when we look upon tone naively as a whole, it is as mass or volume that it appears before us. This implies that it has at least one other dimension, different from that indicated by pitches. The musical aspects of tone give us no means of demonstrating or of defining this direction. For these aspects are concerned only with the variations of tone in pitch and volume, i.e. only with longitudinal variations. And no transverse variation accompanies these. We must turn to quite a different function of hearing, one that is dependent not on either ear separately or equally, but on both ears integratively or in a combined purpose. This function enables us to localise sounds towards the right or the left ear and in an imperfect way round the head in space. A careful study of binaural hearing leads to the conclusion that every tone has breadth as well as length. This breadth is also marked out into a series by the variations of binaural localisations or 'local signs'; and, unlike the pitch series, this one is traversable as far towards the one end ('opposite the one ear') as towards the other ('opposite the other ear').

We have no means of comparing or of measuring the extents of the two series with one another, such as superposition or the like. But we do not feel this ignorance or disability as a difficulty or mystery; for we simply do not think, we have no inclination to think, of these series in relation to one another. On the contrary, the one is the musical aspect of tone, the other is the basis of its spatial aspect. Music is the same for a person of normal hearing whether it is played to the right or to the left of him. And the tonal nature of a warning signal is largely insignificant, if we but gauge the position of its source aright. These two interests of sound appeal to very different practical functions in spite of the close relation of their ultimate bases to one another. Besides, the total transverse breadth of tone probably hardly ever varies, unless in passing from purely uniaural to binaural hearing. It is only the point of emphasis in the total breadth that varies and so forms a basis for localisation towards the one ear or the other. The longitudinal or musical aspects of tone would, therefore, be unaffected in any way. And in case there might be any diversion of interest, we neither encourage our musicians to rove around us while performing, nor do we practise an oscillation between hearing in the usual way and hearing with one ear only. But even though the breadth of tone is not, and from the nature of the case cannot be, a musical variant, it is still there all the time in musical sound, and doubtless makes it what it is—a volume. (For proof of this transverse aspect of sound, see 77, Chap. IX.)

We may sum up our conclusions by saying that pure tone is a volume of sound in which a minute part or point is most intense, while the rest is graded smoothly and symmetrically around this—probably central—part, the pitch of the tone. By means of pitch, tones can be arranged in a series of diminishing volumes. But no two simultaneous tones fail to fuse or blend with one another or to be heard through one another. We therefore infer that the volume of any tone includes not only the pitch, but the whole volume of every higher tone; so that all tones may be reduced to a single series of hypothetical particles of sound, one half of which series we actually hear as the pitches of tones. Apart from these pitch points the nearest approach to the hypothetical particle of sound that is ever separately experienced by us is the minute volume of the highest audible tone or sound. The other dimension of tonal volume appears in the 'local signs' of binaural hearing. But there is no method whereby we might compare the size or length of this series with the length of the pitch series, so as to say what the exact shape of tonal volume is. We must be content with the close correspondence shown between the feeling of tonal volume and what we have proved it to be. The probable shape of tone is like a visual parallelogram that is longer than it is broad. It varies greatly in its length but probably not, or hardly, at all in its breadth. This relation between breadth and length constitutes the typical and constant form of tonal volume. It may be illustrated diagrammatically thus:

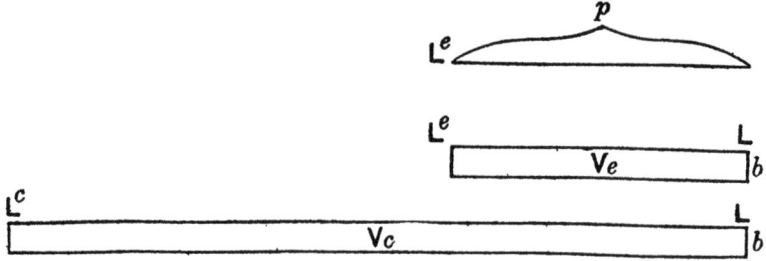

Fig. 2. Diagram of two tones a tenth apart, e.g. c^1 and e^2, showing the constant breadth (b) and the variable lengths (L^e–L and L^c–L) of the volumes of the tones (Vc and Ve), and the intensive differences within the volume (L^e), the pitch (p) being the predominant point of the whole.

CHAPTER III

DEGREES AND THEORY OF CONSONANCE AND
DISSONANCE (FUSION)

In a general way the various grades of consonance between two tones sounded simultaneously have been long established and are disputed by no one. The ancient Greeks recognised three grades of consonance or symphony—the octave, the fifth, and the fourth. To these, in a certain sense[1], the third (but not the sixth) was added by Gaudentius about the end of the third, or the beginning of the fourth, century A.D. (cf. 66, 72). Modern musical theory grades the consonances as perfect—octave, fifth, and fourth, and imperfect—the major and minor thirds and sixths. All the other intervals smaller than the octave are dissonant—the sevenths, the tritone (diminished fifth or augmented fourth) and the seconds. The increase of any interval by an octave is not held to make any change in its consonance or dissonance (except, in order to satisfy the needs of certain theories, in certain chords).

Of the dissonances, however, the minor seventh, as also sometimes the tritone, is commonly considered to be almost consonant. Frequent reference is made (especially by those who look to the series of partials for the causes or origins of chords and scales) to the approximate equality between the diminished seventh and the 'natural' seventh formed between the fourth and the seventh harmonics of any fundamental (in the case of the tritone the fifth and seventh harmonics). The difference is only a sixty-fourth part—in terms of the ratios of vibration of the two partials. It has often been claimed that the chord *cegbb*, taken exactly in the ratios 4, 5, 6, 7, is really a concord (cf. 66, 71). Those who look for some fundamental development, say of the ear itself, underlying the progress of musical art, might claim this tendency as another step forwards beyond the one first recorded by Gaudentius. Various theorists have even believed that all the grades of consonance have their ground in habit, racial if not individual. For the partial that occurs oftenest and loudest with any fundamental is probably its octave, the next is the fifth above, and so on in the series of partials. What we hear oftenest together, the mind (individually) or

---

[1] Which we shall have occasion to examine more closely later on.

the ear (racially) comes to hear as one or at least as an agreeable combination—because the ear or the mind has got accustomed to hearing them together. Thus in its progress the art of music is gradually climbing the ladder of partials. We have already accommodated 1, 2, 3, 4, 5, and 6 in all possible groupings and we are now absorbing 7. In some future we shall bring even 9, and 11, and 13 and others that we now treat as dissonances, over the border as consonances[1]. One writer has even tried to create an experimental basis for this theory by giving persons much practice in placid attention to dissonances.

A number of experimental investigations have been made with the purpose of grading the diatonic intervals more accurately than the usage of music indicates. The general result of these yields the series: octave, fifth, fourth, major third, minor third, major sixth, minor sixth, tritone, major second, minor seventh, minor second, major seventh; or in a useful symbolism, which will be maintained in the following pages, 0, 5, 4, III, 3, VI, 6, T, II, 7, 2, VII. This series represents comparatively to what degree the two tones seem to fuse with one another to form a unitary whole. 0 is more fused than 5, 5 than 4, and so on.

In other experiments an attempt has been made to obtain figures representative of the degree of difference between the grades of fusion. Highly trained musical ears might attempt to indicate these quantities by direct estimation. But their judgments are quite open to the influence of their knowledge of musical practice or of any other thoughts or theories they may have. For they recognise the interval as soon as it is sounded and so can think of it whatever occurs to them. And the relative quantity of fusion is not, as it were, marked on the face of intervals. More or less unmusical minds, however, are often quite unable to distinguish the two pitches of an interval, or to recognise one of them or the interval as a whole, etc. Consequently they think nothing about them by way of memory, so that they are sure to be freer from suggestive influences than musical minds. This advantage

---

[1] Cf. A. E. Hull, 22, 265: "The standard of aesthetics varies from age to age. A combination of notes which one generation accepts only on sufferance will be received by a later generation with equanimity or even delight: Monteverde's Sevenths, Wagner's Ninths, Gounod's Thirteenths, Debussy's Twelfths, and so on." On page 115 Hull gives a scheme of development which places the first two partials as primeval, the second two as mediaeval, 5, 6, 7, 8 and 9 as of the 18th and 19th centuries, 10 to 16 in the form of $e, f\sharp, g, a\flat, b\flat, b$ and c as "whole-tone scale, Debussy, Scriabin." "Undoubtedly Scriabin's exploitation of the higher harmonics will lead to wonderful developments, which are even already in evidence" (p. 271). But he says elsewhere (p. 265), "Many passages in Scriabin's work seem ugly to us, some almos⁸ repulsively so." Cf. 24.

weighs against their want of practice and skill, which can even be turned to special account. For the unmusical mind's failure to detect the pitches in intervals may be used to obtain an index of their grades of fusion. The oftener one fails to detect that two sounds have been presented, the more unitary and fused we may suppose the whole sound mass to have been; the oftener the listener feels that two sounds have been given, the less unitary and fused must the combined sound have appeared to him. One series of experiments in this manner gave 80 % of answers asserting the presence of one tone where there had really been an octave; 50 % where there had been a fifth; 35 % for the fourth; 30 % for the minor third; 27 % for the major third; 23 % for the tritone. These and other results have shown that there is a marked difference between the octave and the fifth, and between the fifth and all the others; but that amongst these last there are at the most only slight differences of quantity. If enough tests are made carefully the usual grading will emerge on the average. But just as in the tests for grading without respect to quantity, the single tests here show frequent reversal of the results that appear on the average. This, of course, only emphasises the approximate equality of the lower grades of fusion.

Those who are not familiar with the notion of fusion must be careful to avoid misapprehension. The results do not suggest that intervals such as the fourth and the thirds are hardly distinguishable. Fusion does not apply to the peculiar aspect of a pair of tones that we call their interval; but to the whole mass of sound formed by the two tones in so far as they merge into, or blend with, one another, or in so far as they appear to be one to a person who does not recognise them as one interval or another, and yet, of course, hears them as a whole. Nor do these grades of fusion imply that the musician does not know and hear the fourth as a greater consonance than the third, or the third than the second. The musician is highly practised; he has learnt by long experience, from tradition, and by harmonic usage to classify the pairs of tones as distinct grades of consonances in spite of the slight differences that may distinguish them, so that these differences perhaps seem greater and more decisive to him than they really are. This distinctness of grading is very much increased by the sharp differences there are between pairs of tones as intervals, whereby they are at once recognisable and distinguishable. For whatever is musically associated with a certain pair of tones can be attached to their distinctive nature as intervals and so can be unfailingly recalled; whereas if it had to be recalled by their indistinctive nature as fusions alone, the slightness of the difference between the lower grades of fusion would make recall very uncertain and liable to confusion, so that in turn the want of distinction amongst the lower grades would be thrown into prominence. In observing fusion the musician has the special difficulty of abstracting from his highly trained knowledge and from the intervallic aspect of tonal pairs. The Greeks, of course, could easily distinguish thirds from seconds as intervals, but it took them long to compare the members of their class of discordant intervals with one another so carefully as to see that the third stood very high in the class and had so much fusion in it that they could place it next to the fourth and even class it in a certain sense as a consonance. Advance in observation and analytic abstraction seems to account for their change of classification more easily than the hypothesis of development does. Improvements in the method of tuning may also have been of influence. But there is, as we shall see, no doubt that the differences between all the intervals with which we are so familiar nowadays has only been fully displayed by the functions they have acquired in our highly developed music.

The explanation of fusion in the first place follows closely the suggestions given by the description of its highest grades. Fusion is degree of resemblance to the unity or balance of a single tone. Two tones that fuse, blend with one another so as to appear more or less evenly intermingled, whether the hearer fails to recognise their difference or whether his natural aptitude and his practice enable him to recognise them at once. The fusion is not altered by the ability to recognise the constituent tones in spite of their fusion. To the talented musician who practically never fails to notice both its tones, an octave is still a high grade blend. Nor does his ear contrive to isolate these two tones from one another so that they shall appear to him to be as separate as they are when successive.

Our previous study of tone showed that the whole series of tones from highest to lowest probably consists of one total series of (hypothetical) particles of sound, of which a number always beginning at the same ('high') end of the series enters into any tone sufficient to make up its volume. The highest audible tone requires approximately only the single ('highest') terminal particle, the lowest the whole series. Now when two tones are given together, the volume of the lower must include the volume of the higher and apart from some special marks the two will not be very easily distinguishable. The coincidence of the two volumes would probably of itself make the upper part of the total volume more intense. And it is even conceivable that the upper volume might thereby be recognisable as such, so that the listener could say : in this total sound there is besides the total extent of sound, a sound of a certain smaller extent. And it may be supposed that these sounds would appear even and smooth (or specifically tonal) in so far as other and irregular changes of intensity are absent in the two volumes.

From this point onwards the attempt might be made to construe the whole nature of tones and their combinations without the use of pitch in the sense, above expounded, of a central point of prominence in the tone's volume. I shall not argue the attempt out in detail. It will suffice to acknowledge its logical possibility here. The use made of the notion of pitch in the following pages will of itself exclude the real possibility of doing without it.

Pitch, as above shown, probably occupies a central position in tone and makes it the balanced symmetrical system it is. If this is so, the overlapping of volumes will yield a special case when the higher volume lies exactly between the pitch of the lower volume and their common higher terminal (cf. fig. 7, p. 199, c and c^1; or fig. 8, p. 200). The whole volume thus constituted will differ from the isolated volume of the lower tone only in the extra intensity of the upper half and in the point of predominance that is the pitch of the higher tone. In so far as this pitch is detectable, the upper tone would be as precisely definable as in isolation. And in so far as the lower end of the upper tone can be felt to coincide exactly with the pitch of the lower tone, or at least to deviate from it when it does not so coincide, this particular case of simultaneous tones would be very precisely definable. And the whole sound would approximate more nearly to the nature of a single pure tone than would any other form of coincidence of volumes and pitches. These things justify completely the identification of the octave fusion with this special case of overlapping.

A second special case would be given when the two special defining points of the higher volume lay equally far away on either side from the predominating point of the lower tone (cf. fig. 7, p. 199, c and g). This case would give a lesser approximation to the balance of the pure tone than does the octave. For in it two new points or breaks of the smooth continuity of the lower volume have been introduced by the presence of the higher. And so the whole would be more easily distinguishable from the pure tone or from one tone than would the octave, whether the two pitches were recognised in the whole or not. By the unpractised person the two pitches would be more easily recognisable than they are in the octave; for the musical mind, there would be less interpenetration of the two tones although he might not be able to detect any difference in the ease with which the two tones were recognised; to both there would be less balance or smoothness in the whole sound than in the case of the octave. We may therefore identify this case with the second grade of fusion—the fifth.

And we are confirmed in so doing by the well known difference between the octave and fifth—namely that octaves are simultaneously compatible with one another, whereas fifths are not. If we add a double octave to the first, the third pitch-point falls exactly into the upper half of the middle tone without disturbing the relative balance of the two lower tones; but if we add a second fifth, it will not thus fit in with the first two tones. On the contrary, it entirely spoils the balance and symmetry of the first fifth.

It follows from the nature of the balance claimed for the fifth that the volumes of its two tones are as 3 to 2. And along with the octave case this implies that the volumes of all tones that have the ordinary fusional relations to one another, are inversely proportional to the corresponding rates of physical vibration. But the relations of the volumes have been educed without any appeal to this familiar physical fact, at least, logically. So long as this is so, it is a matter of impossible speculation to discover whether, given the proper analytic approach to the problem, the volumic ratios would have been discovered and sufficiently proved, had the physical knowledge not preceded. Our only concern need be whether sufficient ground now exists upon which to raise a logically independent proof. This undertaking is not a piece of mere pedantry, as some might think who recall that one way of proving a thing is enough. For the proof given has not for its object the re-proving of truths concerning physical ratios, but the demonstration of an entirely new object, namely the ratios of the volumes of tones, solely as they are heard. This object is a psychological one, if you like. It deals with an object that is as different from physical vibrations as blue is from commotions of the ether or as thought is from brain process.

It is as absurd to suppose that the only objects we can study logically and scientifically and convincingly are physical as it would be to suggest that we could not think correctly until we knew the physiology of the brain perfectly. You may reply that of course we think correctly because the brain is there working correctly whether we know of its workings or not. True; and when we prove things properly about tones, the brain doubtless works properly too. But the chief interest for us is to think correctly first. There is, of course, no doubt that the brain must be capable of interacting in some special way with physical sounds if we are to hear sounds and to think about them correctly. But experience is more than brain. And men may yet have to infer something from the study of sounds as they are for us when heard, if they are to understand how the brain acts in connexion with them. Besides, brain and mind or experience, or more specifically, brain and tones or music are two very distinct and different things that no one could possibly confuse with one another, no matter how much they may be dependent upon, and may interact with one another. So each of them must be studied for its own sake and in the special way its peculiar nature and its relation to ourselves make possible. We cannot bring our knowledge of both into harmony until we have attained a complete knowledge of each.

The form of overlapping of the other intervals and the kind of balance between the parts of the volume they constitute may easily be reckoned out from the knowledge we have just gained. In the case of the fourth, the higher volume will be three-quarters of the length of the lower volume. So the lower limit of the upper tone will fall exactly half-way between the lower limit of the lower tone and its pitch-point. And the pitch-point of the upper tone being at its centre will lie three-eighths of the length of the lower tone from its upper terminal and one-eighth from the pitch-point of the lower tone. The whole mass of sound of the two tones will be marked into parts of two, two, one, and three, eighths of its length. There is here some balance in the one half of the whole volume and less in the other. For the major third the parts are 2, 3, 1, and 4 tenths, which seem more irregular.

As we proceed, the disproportion of the parts seems to grow greater and one or other of the points of the higher tone falls nearer and nearer to the pitch of the lower one. In so doing it must, of course, become more noticeable; just as we catch sight of a second visual point the more easily the nearer (within ordinary limits) it lies to another that we notice easily and are attending to. We notice the pitch-point of the lower more easily because it lies in the centre of the whole volume of sound formed by the two tones; and as pitch is always central in tone and we are constantly at work with the pitches of tones in music, so we get into the habit of attending to tones in a central balanced manner and then notice the pitch and volume of the lowest component of a chord most easily. We do so, not only when the chord is an isolated stationary mass of sound, but also in music where each tone is a phase of a voice or part, except in so far as some melodic figure or theme is present that specially makes out another element than the lowest one of the moment.

But a study of the proportions of the parts of the lower fusions shows no great differences between them. That accords well with the slight differences they show for the ear. And there seems to be no ready way in which we might make our conceptual or theoretical treatment of these volumic parts give us a much more exact account of the degrees of fusion of intervals than the ear gives, so as to provide a rule or rapid guide to the ear, as e.g. measurement is for many purposes a guide and control to the eye. That deficiency does not indicate that the volumic theory of fusion is wrong. The agreement the theory shows between concept and sound, speaks only in its favour. Measurement is by no means an infallible guide or a standard of visual art. It is possible, however, that calculation may yet discover more delicate or more adequate ways of representing the volumic basis of fusion, and from following and moulding itself to portray the verdicts of hearing, may gain strength to place itself ahead of the sense and to lead it forwards to unexplored regions where it may dwell with pleasure and where its art may flourish more abundantly. If that is possible, it would be foolish not to cherish the idea merely because theory has rarely, if ever, yet preceded the experiments of musical art. On the contrary we have every reason to hope that theory will yet be as great a support to art as it has come to be to industry. And we shall do well to build hopes for art upon this dream. After all, such theory does not wish to import into art influences and notions entirely extraneous to its matter, as are ratios of vibrations and all merely physical knowledge. At its best such theory is really only a most perfect description of the actual things that the art of music deals with, tones and their groupings. It gives the artist a better comprehension of their real being than he would gain from his ear alone. It is merely the hearing of the ear perfected and purified by the wide attention and analysis of the intellect. If the art of music can turn such work of the intellect to fruitful use, why should it not be allowed to profit thereby? After all men have had ears and eyes since the dawn of time; but they have not always had the minds to make art out of sounds and sights. Their minds have had to grow to the power of that creation.

CHAPTER IV

THE RELATIONS OF FUSION TO BEATS, PARTIALS, AND DIFFERENCE TONES

HAVING attained what seems to be a satisfactory explanation of fusion, one that is grounded upon the fusing tones themselves and does not involve any reference to any other phenomena that may accompany the simultaneous occurrence of two tones, we may feel free from any obligation to refute those theories that base their explanation of consonance solely upon such adventitious phenomena. Besides these theories have already been well refuted (67); so that before a theory of consonance by volumic balance had become attainable, the discussion of the problem had yielded the conclusion that the grades of fusion are an immediate characteristic of pairs of simultaneous tones and that the most probable explanation of these grades was to be sought in some form of sensory 'synergy' (64, 214). An analogous case is familiar in vision, where certain colours are found to look well together, while others 'kill' their neighbours. We have no very satisfactory explanation of these peculiar harmonies of vision; but the most probable theory supposes that the physiological processes underlying inharmonious or discordant colours in the retina or in the central areas of the brain that subserve vision, do not work easily together or do not make material for each other, as it were, or predispose each other's functions. The notion of 'synergy' is, then, not a specific theory so much as an attempt to indicate where the basis of a true theory will probably be found. And the volumic theory given above may quite well be looked upon as a solution of this query as to the exact nature of fusional 'synergy.'

This by no means recent (1893) reduction of the field in which the explanation of fusion may be sought seems to be still unknown to many of those who are interested in the theoretical foundations of music[1].

---

[1] Thus, for example, in the year 1917 Shirlaw (60, 481) writes: "The only thing which theorists who have made the harmonic series the principle of chord generation appear to have omitted to do has been to abide by the results of their own theory. Having accepted a fundamental and guiding principle of harmony, they have nevertheless refused to be guided by it, and have virtually abandoned it, or, while still professing to do it homage have vainly attempted to exploit it for their own purposes. The principle of harmony of Zarlino, Descartes, Rameau, Tartini, furnishes us with but a single chord.

It may therefore be profitable to make a short review of the critical part of this work now. In so doing we shall best follow the exposition given by the author of the theory of 'synergy.' We shall thereby not only do justice to an important phase of the development of the science of sound, but we may also reach a greater completeness of scientific outlook. For it is quite possible that once the volumic basis of fusion is given, its nature may be heightened or lessened by these other adjuncts of the fusing intervals, even although they are incapable of producing by themselves the effect of consonance or dissonance.

The most familiar and the most widely accepted theory of consonance is that of Helmholtz. Helmholtz gives, as Stumpf (67, 2) pointed out, "not one, but two different definitions of consonance and dissonance, which are indeed closely interwoven in his exposition, but which really are quite different and apply to different fields." According to Heffernan[1] (19) there are indeed no less than three elements in Helmholtz's explanations of consonance.

The chief definition is: "consonance is a continuous, dissonance an intermittent sensation of tone" (20, 226). The intermittence is here created by the beats which appear when tones of about the same number of vibrations per second occur together. As far as they are audible, the number of beats is equal to the difference between the numbers of the two rates of vibration. These beats may originate from the primary tones (as in the interval of a minor second) or from the upper partials of the primaries (as when the fifth partial e^3 of c^1 beats with the fourth partial f^3 of f^1) or from a partial of the one tone with the other primary (as when c^2 beats with the second partial of b^0); it is a matter of indifference what their source is, so long as they make the binary sound noticeably rough. For they must then certainly help to make dissonances less suitable for pleasure and for the clear exposition

But this ought not to be regarded as a negative result, but as a positive result of the greatest theoretical significance. It is the one fact of supreme importance which this principle has to teach us. This has not yet been realised.... There exists in our harmonic music but a single chord, from which all others are developed. But as the sounds of this harmony are contained in the resonance of musical sound itself, all harmony has its source in a single musical sound. The development of harmony has been a more simple and beautiful process than musicians and theorists have imagined." And in a footnote to the word "developed" Shirlaw promises "a new and smaller constructive work on the theory of harmony."

[1] This paper (1887) gives a striking criticism of Helmholtz, but its experimental basis is at times insufficient and lacking in precision.

that art requires than they would otherwise be. And if beatless sounds are of themselves smooth, contrast with the roughness of beating chords will make them seem smoother and more beautiful still.

But it is important to notice that Helmholtz's theory can by no means justify the ascription of smoothness to beatless chords. Smoothness is not a mere negation; a table is not smooth to touch as soon as it ceases to be rough; it is smooth only so long as it gives the finger a continuous area of unvaried sensation. And the regularity and positive smoothness we have shown to exist in tone make any such purely negative use of smoothness inadmissible. Still it is evident that the roughness of beats will heighten and increase the general effect of the asymmetry and unbalanced disproportion of parts that we have shown to constitute the primary being of dissonance. It will make differences and grades of unsuitability and unpleasantness amongst the lower grades of fusion that might otherwise be less distinctive.

So far the theory would thus account only for grades of unpleasantness amongst combinations of simultaneous tones. For successive tones the explanation fails altogether, since no beats then appear in any case. Here Helmholtz appeals to his second basis of consonance in the partials that are common to the successive tones. In the octave we hear again the largest possible part of what we heard before in the lower tone, namely every even numbered partial. In the fifth we hear a lesser part, but still much, namely every partial that is a multiple of three. And so on (20, 253 ff.). We do not need to have analysed these partials from the whole tone by special attention and to know of them. The second tone merely appears to be like the first one according to the amount of identity amongst the partials of the two, as two faces often look similar without our being able to say in what respect precisely they are so.

This second theory is in a notable respect the obverse of the first, as it were. For now the positive quality belongs to consonance alone. Dissonance is a mere negation of consonance; two tones are dissonant when they have nothing in common; or dissonance is the lowest degree of consonance and yet it is not consonance at all. Moreover it is clear that the continuity thus established between successive tones would be a real influence connecting them, in so far as it could be felt in the aggregate; and though it would necessarily of itself be only a weak bond between tones, it would undoubtedly come to the good of any obvious bond already linking consonant tones to one another.

But Helmholtz's second theory does not apply to simultaneous tones. For the partials that are common to two tones an octave apart could at the most only constitute a set of further primary tones to the two from which they originate. Of course they do not under ordinary circumstances do anything of the sort. Besides, Helmholtz's theory is generally quite unable to explain satisfactorily why a tone and its partials are heard as a unit and not as a number of tone spots or separate tones.

"We have thus in fact two different principles in Helmholtz's theory, the one valid exclusively for simultaneous, the other exclusively for successive tones. This state of things seems strangely to have escaped his notice, and as he himself nowhere emphasised the twofold nature of his definition, it has also generally and from the first not been felt to be a defect" (67, 4; cf. 34, 160). Such a duplication and crossing of explanatory principles is certainly derogatory to any theory of the system of consonances and kindred relations in successive and simultaneous tonal groups. That system nowhere suggests any such dual nature; and we could hardly expect two such different causes to yield so homogeneous a system of phenomena. In face of the positive theory of consonance given above these logical and phenomenal inconsistencies would be enough to refute Helmholtz's theories alone. But the following arguments have been stated besides (67, 4 ff.).

As beating is a periodical change of tonal intensity, such intermittence of sensation can easily be produced either in a single tone, or in each of two simultaneous tones. A sounding instrument may, for example, be placed in a closed box from which a tube leads to the surface of a rotating disc; in this disc as many holes as need be are pierced, so that they pass the mouth of the tube when the disc is rotated. Dissonance does not then appear any more than it does when a consonant interval is played as a tremolo. Besides such beats without dissonance, we can have dissonance without beats, as from tuning-forks on their resonance boxes sounding to 500 or 490 and to 700 vibrations, or to 700 and 1000, or to 780 and 1100. Such forks well sounded contain hardly more than a trace of the first partial and at these rates of vibration beats are quite inaudible. Moreover when forks without their resonators, preferably between 800 and 1200 vibrations, are held one before each ear, their tones are not carried to the opposite ears, either by the air or by the bones of the head, so long as they are not made too loud. When two forks of say 800 and 900 (a major tone) are tested in this way, only a trace of beating can be heard at the very most; in striking contrast to the effect of both forks before a single ear; but the dissonance in the two cases remains the same. When in turn a consonance, e.g. a major third (620 and 775 vibrations), is tested thus, it remains just as fused as usual.

The number of beats accompanying any dissonant interval (e.g. a major second) must gradually increase as the interval is raised through the musical range of pitch; but there is no indication that it therefore becomes a dissonance only at a certain pitch or ceases to be a dissonance in the higher octaves. Nevertheless the roughness that is due to beating and that accompanies dissonance under ordinary circumstances will indeed vary with the audibility of the beating (which varies with its rate). Nor do differences of timbre make such regular differences in the degree of dissonance or in the musical usage of intervals as they might be expected to do in view of the different possibilities of beating they create.

But it is needless to follow the argument further. Let us consider the alternative.

The chief argument against consonance by coincidence of partials is its continued appearance amongst tones devoid of partials. And, Helmholtz's work on timbre shows that the different musical instruments, far from having each the complete series of possible partials, differ precisely in the selection from the series that is typical of them; and yet the grades of consonance do not differ from instrument to instrument or from one intonation to another. Suppose, for instance, that the clarinet has, as Helmholtz says[1], only the uneven partials; then an octave on clarinets could not possibly be a consonance at all, but rather an extreme dissonance, because in the second tone we should hear nothing of what we heard in the first. By means of 'interference' we can exclude from a tone any specified partials so as to make coincidence unattainable. Consonance, however, remains unaffected. The consonance of tuning-fork tones is beautiful in spite of the restriction of their partials. Helmholtz may well have been aware of the maintenance of consonance in spite of the relative purity of fork tones. However he may have adjusted his mind to this fact, his successors at least have variously appealed to the influence of memory. But, as Stumpf says (67, 16): "The remembrance that two other blends on the same fundamentals once were consonant, can only bring the

---

[1] According to D. C. Miller's harmonic analyses the seventh, eighth, ninth and tenth partials predominate in the blend of the clarinet. "The seventh partial contains eight per cent. of the total loudness, while the eighth, ninth and tenth contain 18, 15 and 18 per cent. respectively" (40, 201). The second, fourth and sixth partials are very weak.

non-consonance of the present tones by contrast more strongly to my notice. A dish that lacks salt would never be said to be well salted by mere force of memory or custom; on the contrary."

"In short, timbre is for one and the same interval extremely variable, but the degree of consonance is constant. Hence both cannot be explained from one and the same principle. And it is just the happy explanation of timbre that Helmholtz achieved for acoustics for all time that makes his explanation of consonance from the same principle an impossibility" (67, 19).

Various attempts have been made to bring consonance and its grades into relation to another class of phenomena that accompany the simultaneous occurrence of two or more tones, namely difference tones. The earliest such attempt was made by their discoverer[1] Tartini after whom they were often called 'Tartinian tones.' Helmholtz appealed to them in explanation of the less harmonious effect of the minor, as compared with the major, triad, thus introducing a third factor in the creation of consonance. Of recent years an elaborate attempt has been made to base the grades of consonance solely upon the beating and confusion (of a special kind) of neighbouring difference-tones. On the basis of numerous observations of the difference-tones that accompany intervals of different ratios, it was claimed that the greater dissonance had the greater number of difference-tones within close pitch-distance of one another, and would therefore have the more beating and confused blurring. That is to say the latter constitute dissonance[2].

This type of theory seems to escape the criticism fatal to Helmholtz's explanation by coincidence of partials,—that its basis is withdrawn when the primary tones are purified of all partials. For difference-tones still accompany such pure tones. They are not due, like partial tones, to any physical process in the sonorous body or in the air between that and the ear, but they arise somewhere within the ear directly from the

---

[1] Or 'one of their discoverers.' Cf. 60, 301: "Although Tartini is generally regarded as the first to discover the combination tones—he had asserted that as early as 1717 he had made use of them for the purpose of teaching pure intonation on the violin to his pupils—it is certain that other musicians had discovered them independently. J. A. Serre of Geneva, and Romieu of Montpellier, had given accounts of these tones before Tartini's publication of the *Trattato di musica* (1754)." G. A. Sorge in his *Vorgemach der musikalischen Komposition*, "published nine years before Tartini's *Trattato di musica*, demonstrates his acquaintance with the phenomenon of combination tones."

[2] For sources and criticism see 68, 57. For the most trustworthy and complete record of observations of difference-tones see 70.

interaction of the primaries. But difference-tones can be greatly weakened, if not made to disappear entirely when the primary tones are presented one to each ear and are given in somewhat weak strength (cf. 7). The dissonance, however, does not then disappear nor change its degree. Besides it is a fatal defect of this type of theory that it gives no really positive status and explanation to consonance. Consonance is here a mere negation or minimum of dissonance. And whatever be the nature and cause of dissonance that is postulated—whether it consist in the multiplicity of the difference-tones or in the fluctuations of their beating or whether it be traceable rather to the confused indistinguishability of too closely neighbouring difference-tones that form between-tones or even a sort of streak-tone, or the like—we should in any case certainly have no reason to hear non-dissonant sets of tones in any other way than with the greatest precision and clearness of distinction from one another (cf. 68, 282). Even though, as in the great consonances, the number of difference-tones is greatly reduced (to none in the octave), the two primary tones would still be two in all obviousness; there could be no excuse for holding them to be but one, and no ground for establishing any special relation between them (such as that of 'consonance') except that of clear distinguishability. Thus the appeal to difference-tones can only give a partial explanation and must therefore be unsatisfactory.

If, however, a positive explanation of consonance and dissonance in their grades has already been given, as in the previous pages, it is obvious that any beating of difference-tones amongst one another or with the other components of the whole sound would add to the roughness and irregularity inherent in the latter through its primary components, while the consonance of these parts—which would have to rest upon the same kind of processes as the consonance of the primaries,—would bear out the latter. Consider, for example, the case of the perfect fifth in pure tones in relation to the two loudest difference-tones—the 'first' (h–l) and the 'second' ($2l$–h). The ratio for the fifth is $2 : 3$. Its difference-tones are both of ratio 1. A slight mistuning will yield one difference-tone just less than 1 and another just more than 1. These two will beat with one another, whereas in the just interval we shall have three consonant intervals, octave, fifth and twelfth. Similarly in the two common triads, major and minor, whose ratios are $4 : 5 : 6$ and $10 : 12 : 15$ respectively, we find the following components: in the major chord, 1 (twice), 2 (twice), 3, 4, *4*, *5*, *6*, or C, c, g, c^1, e^1, g^1; in the minor chord, 2, 3, 5 (twice), 8, 9, *10*, *12*, *15*, or

$A\flat$, $E\flat$, c, $a\flat$, $b\flat$, c^1, $e^1\flat$, g^1. Apart from the difference of octaves there are three dissonant intervals in the latter chord and fewer high grade consonances (3 octaves, 6 fifths, and 2 fourths to 4 octaves, 5 fifths, and 1 fourth). We shall return to this topic again (p. 192 f.).

The general series of the upper partials and the difference-tones were on the whole a relatively late discovery in the history of the scientific foundations of music. But many years before Helmholtz propounded his very convincing theory of instrumental tone-blend (timbre), they had become familiar to all the leading theorists. If any feature at all of tones were really explicable in terms of some such adventitious accompaniments of primary tones, or rather consisted of them, we might certainly have expected Helmholtz's predecessors to have learned how to explain pitch-blend by the grouping of partials. That they did not do so, and obviously were not tempted to do so, gives us the right to consider it highly improbable, apart from all other grounds, that so direct and unmistakable phenomena as those of consonance and dissonance are founded upon such remote accompaniments of primaries as partials and difference-tones and their beatings. We must find the basis of consonance and dissonance, as it were, directly in or below the primaries themselves. And that the theory propounded above has succeeded in doing.

CHAPTER V

THE CONSONANCE OF SUCCESSIVE TONES

WE have not yet given an account of the consonance of successive tones from the standpoint of the volumic theory. But it is evident that the task is a very simple one and involves no change of the basis of explanation and no new principle. This necessity characterises all the other theories we have noticed. There is no beating between successive tones, so that there can be no roughness between them. And while one tone may certainly repeat a number of the partials of a preceding one, yet there is no means of detecting which of the partials appearing with two simultaneous primaries belong to either, in so far at least as they might belong to both. Finally, the difference-tones that appear with simultaneous tones are lacking in their sequence.

The attempt has been made to cover over these lapses of the basis of explanation by appeal to the restorative work of memory. The idea is that when the basis of consonance or dissonance is actually given, the memory will mark it well and associate it with the primary tones which it accompanies. Later when these primaries appear without the basis of their fusion, this characteristic will be restored by the memory and the primaries will function as fused. There is no general psychological fault in this theory so far as the memory's activity is concerned. Seeing a man often and hearing him speak, we learn to connect his voice with his visual appearance; when we later merely see him we can call his voice vividly to mind.

But we do not then hear him speak. Memory does not cause hallucinations in the most of us, nor should we desire it to do so. There are, however, cases of a less abnormal character which seem to imply a restoration of sensation by memory. Thus a glowing iron is often said to look hot, a child's cheek looks soft and tender, the ground after rain looks wet. True; but these things do not then *feel* hot or tender or wet; they merely look so, because their visual appearance makes us think at once of the associated character that comes through the other sense. There is a certain visual feature in each which prompts the mind to recall the associate, and so that visual feature acquires in our minds a special meaning as a sign of the associate. But the visual 'ground' no more feels wet because it looks so, than the word 'lead' acquires a weight because it is the sign of a heavy thing. On the contrary when a thing looks heavy, it usually feels lighter than another thing of the same weight that does not look heavy or that looks light. This is the size-weight illusion, demonstrable with two objects of equal weight but of different size. Expecting weight we do not feel it to be greater than usual, but less.

At least as much as this is also true of tones in relation to consonance. If you hear c and then d and recognise the interval between them specifically as a whole tone or merely as much smaller than the consonant intervals, you may certainly recall the fact that these tones together would form a dissonance. But they would not therefore sound dissonant in sequence. On the contrary, if you had learnt from the simultaneous tones to expect a dissonance, then on hearing them in sequence you would be greatly struck by the absence of dissonance; just as you would be astonished by the weight of a cigarette you had picked from a box if it happened to be made of lead.

The suggested explanation of fusional degrees by 'synergy' has met with a similar difficulty in explaining the relations of successive tones to consonance. A solution by presumption offers itself readily enough, however. For we may suppose that the special function by which two tones make each other's action easier or harder when simultaneous, still exists when they are successive. For the earlier tone is not then entirely gone, any more than the earlier part of a melodic phrase is mentally non-existent when the later notes are being played. The function of fusion would then hold between tones that are 'together in the mind,' so to speak, whether simultaneously or successively. In other words the mind's sphere of immediate activity, unaided by memory, covers not only the present instant 'now,' but a short reach or length of time. Of course, we should still have to explain why in sequence tones do not fuse in the same way as when simultaneous.

This way of accounting for the relations of consonance both to simultaneity and succession of tones by the same principles seems easy only because no definite theory has been advanced. Only the formal requirements of a successful explanation have been sketched. The other theories, such as Helmholtz's, failed decisively because they claimed to have found a definite cause of consonance or dissonance which criticism has shown to have apparent validity only in respect of simultaneous or successive consonance and to be obviously inapplicable to the complementary case. And Stumpf did not feel quite satisfied with this extension of his notion of 'synergy'; for he suggested various means whereby the relations of successive tones might be brought into closer parallel with the fusion of simultaneous ones, including the principle of relationship through partials, advocated by Helmholtz (cf. 19, 58 ff.). Coincident partials may well give another kind of bond between tones, but that cannot be the bond of consonance, if consonance is to be explained by 'synergy.' We need not debate these notions further now. Let us rather consider the problem from the volumic point of view.

It is immediately clear that two tones an octave apart in sequence must have a special relation to one another as volumes. The higher one will fall in the tonal field exactly upon the upper half of the volume that formed the lower tone. Or the lower will occupy just twice the volume occupied by the previous higher tone; its pitch-point will be exactly where the lower end of the volume of the higher tone lay. When the tones are simultaneous, we notice how perfectly the two fit together to form a regular whole. Perhaps we get our impression here more as a whole than from an analytic study of the coincidence of points. The coincidence is there, of course; but we probably feel the fit as a whole rather than see it or inspect it point for point. When tones follow one another, however, this analytic procedure becomes more possible. We could not, of course, state in exact conceptual terms our procedure in observing the tones, so as to corroborate precisely the theory of their volumes. But the different ways we use our attention might really correspond to the statements we deduce from the theory of the volumes of tones for all that.

Thus, e.g. in noticing the lower tone after a few trials we may well fixate the pitch of it exactly and observe then whether the lower end of the higher tone just touches off that pitch-point. In vision we can describe the procedure of the attention in very precise conceptual terms. We take one line and let the end point of it fall exactly on the end point of another line and make a second point of the line fall on some other point and so on. We are unable to do this in hearing, not because the stuff of sound would not allow of it, nor because our minds are somehow befogged in dealing with tone, but simply because we cannot turn and move tones about in the auditory field as we move figures in the visual field. Nor can we dot any required point into a sound volume as we do with visual lines, and so on.

Similarly in the fifth we pass from the one tone to the other by an easy path. We could not fail to notice the symmetrical relation of the new defining points of the higher volume to the pitch of the lower, even if it were quite impossible to sound two tones at once.

But it is obvious that there is a considerable difference between simultaneity and succession. The former creates a balanced, unitary mass that differs from other such masses in its degree of balance and unity. Sequence creates a passage that may be regarded in much the same way. The lower tone gives way to its octave gracefully, as it were; it almost introduces it, pointing in a sense to the place where it will appear, or preparing a place for it against its coming. The same is true in a different manner for the fifth.

At the same time it is quite possible for the mind's eye to take the measure of the two tones in volumic projection upon one another, as it were, and to see their volumes against one another as if they were simultaneous, without, of course, being so. That is, we can, if we will, take a sequence of tones as if they were a fusion of simultaneous tones and judge them accordingly. The two tones do not, of course, actually fuse; but they have to be taken or heard together as if the first one were still there when the second appears; their intensities are not summated as in the case of simultaneous fusion (cf. p. 52 f. below), but the balance and symmetry of their relative positions are noted. This is done regularly in music in the arpeggio forms of chords. But it is not necessary on the other hand that the mind should always do so. There is not only no reason why it should, but it is easy for it to do otherwise.

Special interests of music, especially the melodic, lead us to take successive tones specifically as a sequence. Here, on the contrary, the tones are apprehended specially as a sequence; we let the first one go and pass from it to the second. In the arpeggio chord we have a whole given successively; in the melody a sequence or motion is given successively. Or in the chord the successive tones are held in projection upon one another, while in the melody they are each complete stages of a transition. For this purpose smallness of interval is a favouring factor; it makes for continuity or for melodic progression[1]. Continuity is present even with the larger interval, but it is then not so obvious or so obtrusive; it may have to be supported by other relations which bring successive tones into connexion with one another. In this way

---

[1] Cf. 41, 246: "In folk music generally the frequency with which the various intervals are used decreases proportionately with their size." It does so also in the melodies of Schubert's songs, as I have determined by sampling every tenth song. Only the minor second occurs less frequently than the major. The figures of the several frequencies are: 2—1673, II—2171, 3—926, III—455, 4—633, T—60, 5—195, 6—118, VI—58, 7—17, VII—0, 0—28, 9—1, IX—0, 10—1, X—2. The sample consisted of 56 songs.

we very often find the melodic and the consonantal aspects conjoined in the same interval. Large intervals enter into melodies more easily when they are such as would be consonances with simultaneous tones (cf. 52, 22)[1]. Not that the greatest consonance—the octave—is oftenest used, the fifth next, and so on. Each interval has to be judged on its merits. The great consonance of the octave is weighed down by the large melodic step required by it and is probably less often used for that reason. The fifth with a lesser consonance will very likely be used oftener because of the greater advantage given by its much smaller step. The matter has not been fully treated statistically, as far as I am aware, but it would probably well repay the trouble necessary to gather the facts.

Thus it seems that in the volumic theory a basis is presented from which all the interests of music in simultaneous and successive tones may be fully satisfied without neglect of any of the differences involved in these two cases.

[1] This is confirmed in the statistics of Schubert's songs.

CHAPTER VI

THE NATURE OF INTERVAL

THE nature of interval has always been one of the great mysteries of sound. It formed for the ancient world a fitting parallel in sense to the wonderful relations shown by numbers. The discovery of the connexion between the grades of consonance and the ratios of the smaller numbers let loose a flood of mysticism which endured for centuries. Rameau seems even to have thought that a thorough explanation of the sensory basis of tonal proportion might lead to an insight into the being of proportion in general and in particular as it appears to us in numbers[1]. He was sharply criticised for this by the Academie des Sciences to whom he presented his scientific plans for approval and support. And their censure of his mystical vanities was re-voiced by D'Alembert[2] in spite of the admiration which Rameau's efforts and success in forming a systematic whole out of the empirical musical wisdom of his time aroused in him. Rameau, of course, did not succeed in explaining the mystery of interval and its relation to

---

[1] 53, 2: "Ne connoissant point la nature de notre Ame, nous ne pouvons apprétier les rapports qui se trouvent entre les différens sentimens dont nous sommes affectés: cependant lorsqu'il s'agit des Sons, nous supposons qu'ils ont entr'eux les mêmes rapports qu'ont entr'elles les causes qui les produisent." "Ce qu'on a dit des Corps sonores doit s'entendre également des Fibres qui tapissent le fond de la Conque de l'Oreille; ces Fibres sont autant de corps sonores auxquels l'Air transmet ses vibrations, et d'où le sentiment des Sons et de l'Harmonie est porté jusqu'à l'Ame" (p. 7). "On peut dire même que la Musique a cet avantage singulier, qu'elle peut toujours offrir en même-tems à l'esprit et aux sens tous les rapports possibles par le moïen d'un Corps sonore mis en mouvement; au lieu que dans les autres parties des Mathématiques l'esprit n'est pas ordinairement aidé par les sens pour appercevoir ces rapports" (Epitre).

[2] "Le corps sonore ne nous donne et ne peut nous donner par lui-même aucune idée des proportions....3°. (et c'est ici la raison principale) parce que, quand on entendrait ces octaves et ces sons des multiples, le sens de l'ouïe ne peut en aucune maniere nous donner la notion de *rapport* et de proportion, que nous ne pouvons acquerir que par la vûe, et par le toucher. Pour avoir une idée nette des proportions et des rapports, il est nécessaire de comparer les corps par ces deux derniers sens; la perception des sons n'y contribue absolument en rien, n'y ajoute rien, y est totalement étrangere. Pour tout dire en un mot, quand les hommes seraient sourds, il n'y en aurait pas moins pour eux, des *rapports*, des *proportions*, une géométrie. En voila, Monsieur [Rameau], plus qu'il n'en faut sur ce sujet; et les Mathématiciens trouveront à coup sûr que j'en ai encore trop dit" (9, 213 f.).

physical ratios. But he was certainly right in feeling that there was something in the sensory experience to be explained which would tell us how we feel proportion in one instance at least, whether that case can throw any light upon other forms of mentally grasped proportion or not. He did well to linger longingly upon the wondrous problem. And D'Alembert's denial was somewhat too sweeping, at least so far as the presence of proportion in hearing is concerned.

That we detect very special and precise features in our tonal experiences correlated to certain very definite proportions in their physical stimulus, should prevent a cautious, logical mind from asserting point blank that proportion has absolutely no place in hearing.

D'Alembert, like so many others since his day, was convinced that the 'metaphysics' or psychology of hearing would "according to all appearances always remain covered with clouds." And yet he somehow convinced himself at the instigation of Rameau's researches that the principal laws of harmony could be deduced from a single experiment (9, xxvii). But we now know that that idea is really as absurd as Rameau's speculations on proportion seemed to D'Alembert himself. In fact it is worse. For Rameau did include the phenomena of hearing in his field of search, whereas D'Alembert seems to have thought that a physical experiment or relation was worthy of the place of honour at the feast of music without wearing the garment of experience. Criticism has since thrown that and all other intruders out into the limbo they belong to. And the clouds have blown away.

In fact the solution is not by any means difficult to attain or to apprehend, once the fundamental secret of tone has been discovered and understood. However that may be, the wonder of it all remains that sound and hearing should be so cunningly devised; that the weft of nature's mighty looms should reveal so beautiful a pattern in this auditory part. And the greater wonder still is that our intellect should have been able from this slender basis to raise the great art of music to such complexity. Our task in the following pages will be to try to show how the great edifice of music is placed secure on its foundations and how it is carried upwards towards the art as we know it. No one who follows this science of tone from its beginnings can fail to be struck by the extraordinary nature of sound and the marvellous skill with which music has been created by man.

The oldest theory from which an explanation of interval was sought started from the obvious fact of vibration in the sonorous body and the relation between pitch and rate of vibration. "This doctrine, first taught by the illustrious founder of the sect [Pythagoras], adopted and developed by Lasos, by Aristotle, by Euclid, and later by the neo-platonists, has been formulated by Nicomachus, whose words Boethius transmits to us. 'It is not,' he says, 'a single vibration that produces a uniform sound; but the string, once set in motion, gives birth to numerous sounds, because it impresses frequent vibrations upon the air. But, as the rapidity of these shocks [of the air] is so great that one sound is confounded in some way with the other, we do not perceive the distance [that separates them], and it is as it were a single sound that reaches our ears. Now when the vibrations of the low notes and the high notes are commensurable amongst themselves (as for example in the proportions indicated above), there is no doubt that these common measures blend together and produce the unity of sounds we call consonance'" (14, 96f.; 3, I, 31)[1]. Keeping in touch with the progress of the physical science of sound, this doctrine has been carried down to our own time. It was the basis of Euler's *Tentamen novae theoriae musicae ex certissimis harmoniae principiis*, from which the Table opposite his page 36 has been copied so often. Thus the octave gives a pattern of this kind—: · · : · · :—the upper line of dots representing the waves of the higher tone, the lower the slower waves of the lower tone. The pattern for the fifth would be thus—: ·.· : : ·.· : (2 : 3). Probably the best and at the same time the most self-critical statement this theory has ever received was made by a Scotsman, John Holden, in an Essay towards a rational system of music, published in Glasgow in 1770. The psychological analogies he brought forward are admirable. Even in recent years the theory has been renovated by Th. Lipps, who believed that these waves were carried to the brain and transferred to subconsciousness, there being a unit of process in the latter for each physical wave. Somehow this sequence took on for, or in, consciousness the form of a smooth unity. The rhythmic coincidence of the processes of subconsciousness that went on when two tones were sounded, was supposed to be felt by consciousness as consonance, want of rhythm or its puzzling complexity as dissonance.

---

[1] "Non, inquit, unus tantum pulsus est, qui simplicem modum vocis emittat, sed semel percussus nervus saepius aerem pellens multas efficit voces. Sed quia ea velocitas est percussionis ut sonus sonum quodammodo comprehendat distantia non sentitur, et quasi una vox auribus venit. Si igitur percussiones gravium sonorum commensurabiles sint percussionibus acutorum sonorum, ut in his proportionibus quas suprà retulimus, non est dubium quin ipsa commensuratio sibimet misceatur, unamque vocum efficiat consonantiam."

"It is not," as John Curwen said (8, 10), "that the mind actually *counts* the relative number of vibrations and consciously ascertains that one tone gives exactly half as many as the other. But by one of those rapid though complex mental processes which are the marvel of the philosopher, it *feels the result*," adding afterwards in a similar context, "in a way the Great Creator only knows."

The earliest forms of this theory were a legitimate attempt to bring into connexion the two ends of the psycho-physical process, where knowledge offers itself most readily,—the physical and the auditory. But for later theorists, who realise the gap there is for all systematic possibilities between merely felt grades of consonance and ratios of physical vibration, whether of the air or of the ear, the theory is merely an effort to make bricks of straw. Besides, as Rousseau noticed (56, Art. 'Consonance,' 14th paragraph), how is the mind to catch the rhythm or whatever it may be called, when the periods do not begin and end at the same time, or, as we now say, when their phases are not properly coincident? We shall not spend time discussing any forms of the theory. There is nothing to discuss but mere speculation or ignorance trying to "materialise" itself to knowledge. Ignorance does not breed knowledge; it is the waste land of science to be gradually conquered by the shoots of knowledge that spread into it. Every attempt to bridge the gap between vibration and sound must rest upon greater success in the description of the physical process or of the sounds themselves. For a complete study of either must finally lead to the other, just as one real process binds the two into a single event. In this case our way of success begins from the auditory side.

The line of progress is, in fact, continuous with the theory of tones already developed. It is easily seen that if the upper tone fits so perfectly into the lower tone in the case of the octave, it will do so, however large the volume of the lower tone may be, so long as its volume is the perfect fit. The lower tone may be moved gradually from the lowest reach of the musical range of pitch till the upper tone reaches the opposite extreme. What is common in this series will constitute the interval of the octave as against its fusion or consonance. What is this common feature?

At first glance there seems to be nothing that one can claim as the basis of interval, since the balance or unity of the whole has been allocated to the heard fusion. Even if this allocation was in the first instance the outcome of a process of logical exclusion (77, 60 ff.) it is confirmed by the kinship of the two terms thus brought into connexion —namely heard fusional unity or balance and conceptually formulated balance or unity. A further process of discovery by exclusion seems difficult in the case of the octave, because the fusional aspect of the bi-tonal mass is here so prominent, both for sense and for conception. Let us therefore consider a case from the lower grades of fusion.

There is only a slight difference between the major and the minor thirds or between the different seconds and sevenths in the matter of consonance or dissonance. If these bi-tonal masses had no other feature than their fusion they would never have become so distinctive as they now are in music. Then there must be some other feature in them that provides a basis for our sense of interval.

Let us abstract for a moment from balance and unity altogether, as if we did not apprehend it. Then we may make the following assertion. So long as we were capable of noting the pitches and the volumes of tones, *even supposing they did not overlap* (provided only they consisted of a number of particles or 'spots' of sound, each ordinally distinct and fixed, and so capable of being repeated precisely any number of times), we should still be able to note the relative proportions of their volumes and to construct to any given volume X a volume Y so that their proportion should be the same as that of a standard pair P and Q. We might not be able to do this as well as we judge and reproduce intervals under present circumstances. Our margin of error would probably be greater, just as it is when we compare the lengths or proportions of visual lines from an unusual standpoint. We usually place them directly in front of us and squarely to the line of sight. If we could judge the proportions of volumes under these circumstances, it must be evident that the comparison of the relative volumes of a pair of tones is not made more difficult by the fact that the lower volume consists partly of or includes the volume of the higher tone, so long as the higher volume is distinguishable in the lower. In fact it may well be easier; for the volumes appear in the same place in the auditory field. And the ease of comparison is increased by the facility with which the volumes can be observed in succession.

In thus appealing to a sense of proportion we are merely giving greater scope to a faculty of mind that experimental study has in recent years shown to be of the greatest importance and of the finest efficiency. If a visual standard of proportion is given, say two lines forming the sides of a parallelogram, a fourth line can be constructed to a given third that will show the same proportion with only a very slight error.

The same sort of proportion can be carried through even with intervals of time.

In any case since tonal intervals can, as a matter of fact, be so finely learnt and reproduced as every musician knows, and since there is so strong evidence that tones are volumes of sound of definite magnitudes, consisting of a definite part of a fixed series of auditory particles, each differing from the other only in its place in the series, we have every right to claim that the real basis of our sense of interval is our observation of a constant proportion between the volumes of tones.

This claim, though it has been won by careful theory, that carries our minds through and beyond what the bare tones themselves suggest to our simple observation, is in the end confirmed by our observation. We have only to ask ourselves : is not what we call interval a constant proportion between tones as we hear them? We shall perhaps not assent at once if we merely observe a single interval reflectively. But take that interval and think it successively on to a long series of tones of different pitch. Such a test will show that we are in every way as fully justified in translating sense of interval into sense of proportion as we are in speaking of a sense of proportion in any department of experience at all. We are in the sense of interval finding the proportions of things that really bear proportion to one another, and we do so very accurately.

When we establish relations of proportion between lengths of line by mere visual inspection of them, what do we do? And what are we aware of? We inspect these lines and compare them as to their lengths which appear as sensible magnitudes. We base our judgment of proportion upon this inspection. We are aware of the magnitudes we inspect as lengths and we feel keenly whether a known or given standard of proportion is repeated in a given pair of lines, making thereby in our judgments only a very small margin of error. In judging the proportions of tones, or in judging tonal interval we do exactly the same. We inspect tonal lines of a little breadth, or, as we usually call them, tonal volumes. In these volumes pitches appear, not detracting from our power to judge of interval, but rather aiding it considerably by giving it a sort of focus. We are aware that the tones we compare have volumes or that the whole volume constituting an interval has a particular volumic figure. We are aware of this even though we could not describe what we are aware of in the clear conceptual terms we readily use in vision. For in vision we are all both in practice and in theory highly expert, whereas in hearing most of us are in practice very inexpert and we have all been devoid of proper theoretical insight. Now that the insight has come, we can see that it gives a true description of what we do and of what occupies our attention while we estimate interval. In judging interval we also feel keenly whether a known or given standard of proportion is repeated in a given interval and the margin of error made by expert judges is very small.

The study of these lower grades of consonance as intervals shows us, moreover, that we can fix any interval as an interval in our memory. The interval may be 24 : 31 equally as well as 24 : 32, provided it be fixed in the memory by frequent repetition and attention. Of course it is much easier to learn the consonant intervals because they have a special attraction for the attention and for the memory. For on the one hand they fuse, and on the other they are few and important. Intervals, such as the tritone and the major seventh, which differ only a little in size from some prominent consonance, are hard to sing because they tend to slide into the easy consonance, as it were. But with sufficient practice any such difficulty may be overcome. As Alfred Day said in a similar connexion : "Practice is for the purpose of overcoming difficulties and not of evading them" (10, 7).

It is conceivable, as some have claimed, that those who constantly practise with the intervals of equal temperament should finally come to use them and to think in them by preference[1]. When the circumstances of judging are most favourable, the accuracy with which deviations from a familiar interval can be detected is very great. Thus Meyer and Stumpf got collective results showing *inter alia* an accuracy of 74 % for a deviation of − 0·78 vb. from an ascending major third (600 vbs.); + 2·18 gave 72 %. An individual result (Stumpf's) gave 88 % for − 0·78 from the ascending third and 82 % for + 2·18 (73, 358 ff.).

This process of abstraction has thus yielded us a new feature of complex volumes, namely the proportion of their parts or interval. We may now look back upon the well-balanced fusions that seemed to offer no other feature for analysis than their obvious balance, and reconsider the problem.

The octave, we may say now, is not only a special fusion; it is an interval as well. The tones that form it do not only fit peculiarly into one another, but they also bear a certain volumic proportion to one another. Thus we have a double basis by which to fix the octave in

---

[1] This sentence bears no reference to the controversy on the respective merits of equal and just temperament.

the attention and memory and a double use for it in music. If we ask what are the respective contributions of fusion and of interval to the importance of the role played by the octave in music, there can be no hesitation as to the answer. By far the more important aspect of it is its fusion. Had we not a linear field of sound, but an areal one, as in vision, in which tones could be given at varying distances from one another without overlapping at all, we should have attached as little importance in music to a 1 : 2 proportion between tonal volumes as we attach to that proportion between the lengths of lines in visual art. The 1 : 2 proportion stands forth in music because the 'upper' ends of all tones are identical and tones overlap completely from thence 'downwards.' It is idle to speculate as to what kind of music we might have made if we had had such an areal, or even a cubic, field of sound. I mean, of course, areal for musical purposes. It is areal already, as shown above, as a whole, but the transverse dimension has no musical utility.

Interval might well be called our sense of form in sound, when fusion would be our sense of mass, as it were. There is no use in labouring these analogies between sight and sound, except in so far as they help to bring out the underlying identity of structure in the two senses, and so to understand the nature of each better. Still less should we attempt to base practical reforms or advances upon these interpretations by trying to raise upon the foundations of tonal mass or tonal form structures analogous to those developed in the visual arts upon the foundations of those names. If such structures are naturally possible to music, they will probably have been created to some extent already. If the analogies suggested are real, the first event to follow may well be the discovery that certain types of music differ by their emphasis upon fusion or upon interval, upon mass or upon form.

Possibly that is the real meaning of the great difference of nature and view between harmonic and polyphonic music, the former being the art of mass of fusional (consonantal) effects, of course. No visual art is purely a construction of masses or of forms alone. It is impossible to separate mass and form in this way. Every mass must have a form, and every form that is at least bi-dimensional indicates mass to some degree or other. Visual arts do, nevertheless, differ in the relative extent to which they build on mass and form. Similarly in music. Every fusion has a form and every interval has some degree of balance or mass unity about it. But polyphonic is commonly said to differ from harmonic

music in that the one is viewed horizontally, the other perpendicularly;
or in that the one regards chords rather as a whole, while the other
takes more interest in creating and following out the lines, as it were,
that run side by side throughout the successive groups of sounds. The
difference is one of degree. We shall see as we proceed, that this distinction of attitudes towards groups of tones is of the greatest importance
for a study of the foundations of music.

CHAPTER VII

THE MUSICAL RANGE OF PITCH

ONE of the most curious facts of hearing is that music is restricted to a certain range of pitch. Outside the limits thereof it is no longer possible to make music. The pianoforte makes these limits familiar to every one. The lowest tone on the large concert grand piano is A^2, the highest is c^5. These pitches include a little more than seven octaves. Anyone may notice upon the piano how the lowest notes seem to give an insufficient difference of pitch from their neighbours. The intervals of a major second seem too small; those of a minor second seem to be hardly distinguishable as intervals. A little more, one thinks, and the two tones would seem to be the same. At the upper end of the keyboard a similar change is noticed, though it is not nearly so distinct upon the piano. But it appears clearly if we carry the pitch of tone physically some distance into the c^5 octave.

This limitation of range does not depend upon any purely physical restriction. Periodic waves can be produced below and above these limits and pairs of tones maintain their proper physical relations to one another unchanged. Nor does the phenomenon seem to depend upon an incapacity of the ear to hear tone. For the ear responds with a tone-like sound to physical rates of vibration at least four or five times as great as that required for c^5. A great deal of patient effort and ingenuity has been spent upon the attempt to fix the vibrational limits of hearing as accurately as possible. They will always be uncertain; for their physical sources are not only hard to control and to gauge correctly, but the range of hearing varies considerably from person to person and from youth to age. It also varies considerably with the intensity of the physical stimulus. In fact it is possible that within a large range of physical differences any rate of vibration will produce some auditory effect or other, and if loud enough it may be a tonal effect, without there being any real differences in these effects, except minor or accidental ones. The determination of the limits of hearing would thus be illusory, after a certain point. We shall realise this possibility better further on.

Material has been gathered carefully by experimental means towards an adequate description of the limits of the musical range of tone. It is found that no sharp boundaries mark it out. Towards the upper side it shows itself first in a slight apparent flattening of the pitch of tones from what the rate of vibration of its source leads us to expect. Gradually as the tone is raised this flattening increases to a semitone, and even to a tone. Beyond this point the estimation of pitch soon breaks down altogether. A similar gradual deterioration of judgment is found on the lower side, but here the very low tones seem to be sharper than they should be, according to the known physical rates of vibration.

If interval is constituted by constant proportion of volume, it follows that so long as the pitch of a tone of these high or low regions can be estimated with confidence and regularity, there is at least on the phenomenal side or in the tones themselves nothing amiss. The octave is still in every way an octave for hearing. The discrepancy lies only between the auditory and the physical series. It requires a greater ratio of *physical* vibration to produce a volumic octave in the high border region than is usual in the middle range of hearing. The ratio is $1 : 2 +$, instead of $1 : 2$. Similarly in the low border region the *physical* ratio downwards is $1 : \frac{1}{2} -$ instead of exactly $1 : \frac{1}{2}$. On the higher side there is therefore evidently some difficulty in making the volume of tone smaller in the usual proportion. So the rate of vibration has to be increased a little in order to get a reduction of the volume by half exactly. On the lower side there is evidently a difficulty in making a volume of the usual large size. The sizes required are so great that a reduction of the rate of vibration beyond the half is necessary in order to get precisely the double volume.

These special difficulties receive an easy explanation by reference to the physical sense-organ. The cochlea is quite a small thing, and although its functions seem to be remarkably independent of its size, they can be so only relatively, not absolutely. There must come a point at which the organ will fail to respond properly to a very short wavelength of vibration. Similarly it will be at some point or other finally unable to accommodate the great long waves of sound. No apparatus at all will cover an infinite range of forces. It will fail to work beyond certain extremes, and towards these it will lag behind the change of force applied. At first this will be only a perceptible lag, finally no further change will be given. The ear will respond with the extreme possible to it on either side.

This seems perhaps to be the case at the upper pitch limit of hearing.

For a considerable period beyond the end of the musical range tones are still heard. They become slowly thinner and sharper and finally disappear gradually into a mere hiss or puff. At the lower end the longer waves seem after a time to produce no more effect upon the ear at all. At the most they are felt as puffs of air against the drum of the ear.

The musical range of hearing, then, is the range within which the change of tonal volume keeps march with the change of vibratory rate. As long as this holds good, instruments may be constructed and played with freedom and with complete certainty as to the musical effect upon the listener. It is, of course, conceivable that musical work could be carried up to the outer limits of the border region for a single listener at least. But the physical ratios required at these last points in order to maintain the desired relationships of tonal volume would not be generally valid. They vary considerably from person to person. Consequently music is more or less obliged to discard these border regions in so far as precise effects are desired. Thus the musical range of tone becomes the range within which the changes of volume and of vibratory rate are exactly inversely proportional to one another.

CHAPTER VIII

OUR POINT OF VIEW TOWARDS THE AUDITORY FIELD

WHEN the physics and physiology of vision had advanced far enough to understand roughly the build and functions of the eye, it appeared evident that the impression cast through the lens of the eye upon the sensitive surface was —like that seen on the ground-glass focusing-plate of a camera—inverted. To many men this seemed an extraordinary fact. It yielded for their minds a fundamental problem : to show by what means the image of the eye was turned back to its proper orientation. For we do not see things upside down at all. Consequently either the retinal image must be so transmitted to the brain as to arrive there right side up, or the soul itself must give us a properly adjusted view for the inverted impression it receives. Echoes of this kind of reasoning may be found in books of no distant date.

One philosophic answer to this problem seemed to make it ridiculous. That was the claim that it was here a question not of absolute, but only of relative positions. As all our vision is 'inverted,' none of it is. If the whole world expanded suddenly threefold or if time shrivelled to twice its present rate, we should none of us be aware of the change. So it is a matter of indifference whether visual impressions reach the brain erect or sloped or inverted, so long as they are all modified in the same way. For all we know they may be distorted in the strangest ways in the process of being accommodated to the zig-zag turns of the cerebral convolutions. A certain eminent writer has even pointed out that in our field of vision there is no trace of the holes and slits in the neural continuity that must correspond to blood vessels and connective tissue and such like. He considers this to be an anomalous feature in any systematic co-ordination of brain and mind.

The philosophic ridicule of this problem of inversion is just and proper as far as it goes. In the first instance, or primitively, as it were, it is a matter of indifference how the visual field is orientated, if indeed it can be said to have an orientation to anything outside itself at all. The difficulty of any such absolute orientation may be illustrated in the terms of popular metaphysics. That supposes very often that besides body and mind we have a soul. And perhaps the soul has, or possesses, the mind. In the opinion of some it is the soul that gives us what we find 'in our minds,' that is to say, our experiences. The body somehow acts upon the soul and the soul responds in its own unique and scientifically incomprehensible way. If so, then what is the exact place occupied by the soul? Is it in the brain, or at the brain? The question seems absurd to some. They say the soul has no place. It can even be acted upon from two places at once, e.g. from the two eyes or the two ears, and it then responds by giving us unitary experience. But, nevertheless, for all we know the soul might be far away from the body, say in the star Sirius, so long as an arrangement had been made whereby it should be acted upon by the particular human body on our planet that belongs to it. The problem of the absolute orientation of the visual field is just as insoluble as this problem of the soul's distance from the body. Vision has no absolute orientation to anything that could ever be discovered by us.

But has it not an orientation towards the soul? Suppose you invert the printed page your eyes are now fixed upon, and try to read it. If your soul may not be disconcerted by the change, your faculty of reading will surely feel a difficulty. The disturbance is almost as great when the page is turned through a quarter circle. One of the devices often used in order that differences of colour may be more striking and clear, is to bend down so that the head is inverted and to view the object or the landscape from this unusual position. We may even arrange in this way that nothing in the whole field of vision remains uninverted, or visible in its uninverted relation, not even our own cheeks and eyebrows. Whatever may have been the case in the absolute beginning, there can be no doubt that we, brain or soul or both, do get accustomed to one mode of presentation.

An American psychologist threw much light on this question by wearing for many days in succession glasses that inverted the whole of his visual field. "The first effect was to make things, as seen, appear to be in a totally different place from that in which they were felt. But this discord between visual and tactual positions tended gradually to disappear; not that the visual scene finally turned to the position it had before the inversion, but rather the tactual feeling of things tended to swing into line with the altered sight of them. The observer came more and more to refer his touch impressions to the place where he saw the object to be; so that it was clearly a mere matter of time when a complete agreement of touch and sight would be secured under these unusual conditions. And when once the sight of things and the feeling of them accord perfectly, then all that we mean by upright vision has been attained" (63, 147).

From this important experiment, so trying to the patience of the experimenter, we must infer that, even if the visual field has no absolute orientation, it has at least a correlation with the other sensory fields. Visual 'up' is connected in our minds with muscular 'up,' and so on.

But there must be still more than this. The particles or minimal spots of which any sensory field may properly be held to consist, are both absolutely and relatively different from one another. This difference applies to their ordinal attribute. The particles of sight we call 'up,' are ordinally what they are; they have an absolute differentia inherent in them. This order of theirs we connect by association with a muscular particle (or a series leading thereto) which has also an absolute 'order' of its own. But should circumstances suggest it, we are free to change this connexion by association, so that another visual particle, ordinally very different, will come to be correlated with the muscular 'up.' And so on. Not that these orders are, as it were, absolute places in the universe. But they cannot be called merely relative, because it is not thinking alone that gives them their order towards one another. They come to our thought already ordered; they are already such that if and when we gather them together, we shall see that they actually form a system. That is, their order is inherent in each of them; it is absolute.

But this does not prevent us from thinking these orders in relation to one another and abstracting from their absolute basis or from the associations that rest upon the latter. We can turn a triangle or a square about in the visual field so that it takes up almost any sort of orientation within or upon the absolute constituents of that field. And so we learn to think a square and a triangle independently of its orientation. But if we do not have occasion to make these variations, we shall learn to see a figure and even to recognise it best, or perhaps only, when it is placed a certain way up. Sometimes we cannot easily make these variations. In other cases there are advantages in avoiding them; for one and the same figure—from the point of view of that figure only—may become several figures, if presented in certain fixed orientations and associated with different meanings in each case. Thus h and y (as written) are almost inversions of one another, and yet they are used as signs of different sounds. So are many other pairs of letters. The advantages are here all in favour of letting the absolutist tendencies of visual orientation prevail.

Now all this kind of thing may be in general quite familiar in vision. In hearing, however, where something similar occurs, both the facts and their explanation are probably much less familiar.

The pitch of a chord that is perfectly stationary and is not at the moment apprehended as part of a melodic sequence, is most frequently felt to be the pitch of its lowest component, even when that is not the strongest. The attention seems to fall most easily upon the lowest tone. It may certainly be directed by melodic means or by voluntary effort upon any other component of the whole sound, whether that be a primary tone or an upper partial or a difference tone, or the like. But left to itself and unguided it falls back upon the lowest component, if it is not too weak. It will even fall without instruction upon the lowest difference-tone that may be present.

Amongst the ancient Greeks it appears that the instrumental accompaniment was always above or higher in pitch than the melodic voice. "In the twelfth problem it is explicitly stated that the voice occupies the lower part of the harmony. 'Why does the lower of two notes always take up the melody?[1]...' There are extant in Plutarch two texts no less decisive of which the first is: 'What is the cause of consonance and why, when consonant sounds are struck simultaneously, does the melody belong to the lower?' And the second: 'In the same way as, if two consonant sounds are taken, it is the lower that makes the song.' The custom of putting the accompaniment higher seems to have maintained itself during the Roman period" (14, 364). The accompaniment might descend to unison with the voice, but not go below it. In early Western music the vox principalis was at first higher than the vox organalis, but after a time it took the lower place and remained there (81, 96).

This constant obviousness of the lowest tone is an important factor in music, where, as Macfarren said, the bass "is always the most sonorous part in the harmony" (35, 99). It is to be explained, as already indicated, by the fact that the lowest tone includes all simultaneous higher tones within its volume, and that the pitch of the lowest component is the central point of the whole tonal mass of any moment. Tone is specifically balanced or graded volume of sound. In so far as we apprehend sounds tonally at all, we must look at

---

[1] Instead of "melody" Stumpf understands in the first place "pitch" (65, 19). Compare the suggestions raised by my conclusions below, Chap. XVIII, end. In connected music the more prominent pitch of an isolated interval would become the more prominent melody.

them centrally, as it were. Thus whatever else we may observe in a tonal mass, if we apprehend it as a whole, we shall inevitably look at it centrally and so most readily come upon the pitch of its largest volume, i.e. of the lowest component.

This prevailing attitude shows itself in a number of other ways. We agree in reckoning the interval between any two pitches upwards, unless some special indication to the contrary is given. Thus C–E is to be taken as a major third, not as a minor sixth. Even the Greeks, who in practice found it more fitting to pass from high to low than conversely, reckoned all intervals theoretically from below upwards (16, 89; 14, 173 ff.). Rising of pitch gives us the impression of departure, lowering of pitch that of approach. We incline to take a scale from below upwards and back again, rather than downwards and up again to the starting-point. In a major chord we consider the major third as the first of the two intervals, the minor third as the second. The tonic of a major chord, whether its component tones are given successively or simultaneously, is held to be the lowest of the three. The attempt has been made to look upon the highest tone of the minor triad as its root or tonic. But the very strangeness of the claim, apart from the validity of the special arguments advanced in its favour, shows that it does not correspond to our actual attitude towards the chord. Moreover, when an interval is mistaken for a single tone, as in the experiments on fusion, the pitch ascribed to it is in the majority of cases that of the lower tone.

Another strong evidence of this central attitude to tonal groups is found in certain striking differences between ascending and descending intervals involving the same tones. "The most of those who can recognise intervals at all have learnt in the first place to judge them in the ascending form. The estimation of descending intervals is much harder; in fact it is primarily quite a different task.... The difficulty of judging descending intervals appears not only amongst less practised observers, but also amongst...persons who had all enjoyed a good musical education. In judging descending intervals indirect criteria were often used by them. The time spent in recognising these intervals was also larger than that required for ascending intervals" (37, 192).

This may seem at first to be a very extraordinary fact, hardly creditable by those who recognise all intervals at once without hesitation or by those who find difficulty in naming any by ear alone. It is certainly incompatible with any purely relativistic interpretation. For if the relation a to b is recognisable, the relation of b to a should be so also as a matter of course, since it is the same relation. But if a point of view has been adopted and if a and b are not of a purely qualitative nature, we can readily understand that the appearance of a–b from the standpoint of b may be very different from its appearance from the position a. The face of a friend seems very strange when it is seen inverted. Even the letters of our alphabet or simple ornamental figures become unfamiliar then.

This peculiar difference produced by the direction of interval is seconded by a similar distinction between simultaneous and successive intervals. When musically untrained persons have been taught to attach the correct names to simultaneous intervals, it is found that they are quite incapable of naming the corresponding successive forms, in spite of great practice at the former task. They cannot mentally convert the succession into simultaneity (37, 192).

We may therefore look upon it as well founded that we do adopt a strangely prevailing attitude towards the tonal series. Our point of view is for any moment that of the centre of the whole tonal mass. This centre need not, of course, merely be the centre of a single tone or of the momentary mass of sound that is at the ear. It may be the centre of a tonal complex begun a moment ago and lasting on for a span of time till it is completed. How long this span may be, will depend greatly upon our musical practice and upon the musical coherence and stability of the complex that is presented. These complexes vary in length and complexity very much. It is also possible that in viewing the various parts of this complex as they flow past us, we do not need to maintain the central position that is valid for the whole complex in any rigid way, so long as our disposition towards it remains ready and active. We may then wander about with the centres of each momentary sound mass, always having it in our power to see the relation of that to the general centre of the whole and to return to the latter if required. All this would, of course, not be an inevitable and unshakable consequence of the primary centrality of tone, but would gradually develop out of it by the practice and mental skill of the listener and by the support given to him by the devices of musical art. Thus we see how the primary point of view towards tone might develop towards the special point of view we know in music as tonality, the feeling for a tonic, a point of reference for the tones and chords of a musical unit.

Nor does the fact that the central point of view towards tone is the natural and prevalent attitude prevent us from acquiring another point of view by special practice or preference. Those who sing a certain part in songs or hymns regularly and whose musical practice is predominantly of this kind, will doubtless find it easier to follow their usual habit. The musical analysis of many persons is confined to attention to soprano melodies. Even if their analysis goes beyond this, their greatest practice and interest may yet tend oftenest to the highest voice, so that if a single chord be given they will select from it as its pitch its highest (primary) component. Modern music teaches everyone to pay special attention to soprano melody. For it commonly endeavours to put the maximum of interest into one such melody and subordinates the melodic interest of other voices to their harmonic beauty. D. F. Tovey expresses this when he defines melody as "the surface of music" (75). Our present interest is not to investigate or to depreciate the importance of any such special points of view towards music, but only to show how various facts indicate that the natural, original or fundamental point of view is a central one, or extends from a variable point or centre *upwards* in the tonal range.

At the same time these facts support strongly the theory of the volumic proportional nature of interval and bring the study of tone and music into most intimate agreement with facts that better natural endowment, the greater scope of physical variation and greater practice have made more or less familiar to us all in vision.

CHAPTER IX

THE RELATIVE IMPORTANCE OF SYNTHESIS AND OF ANALYSIS

WE have now gone so far as to be able to look back upon the field of tone and to survey it somewhat as a whole.

The opinion has been often expressed that science can never give a proper account of any art, because the aim of science is analytic; it strives to dissect and to divide, tracing each part to its separate root and origin. It must necessarily lose the life and spirit of the whole. No doubt this is true so long as a science is busy over the preliminary efforts of analysis and has not yet reached the stage of tracing the synthesis that binds the many parts together. But analysis is not the final condition in which scientific results are to be left.

The study of the body involves a long course of special study of each distinguishable part and of its own particular functions. It is only thus that we can learn what primary functions or processes are at work in the living body. And it is only from the basis of this knowledge that we can venture to explain the united work or the integrative action of the living body.

The study of the mind at first calls for a minute examination of every distinguishable experience, its fundamental variability and its primary relations to other experiences. But the science of the mind is not to be taken as a mere catalogue of pieces and processes without connexion with one another. That would be to mistake its achievements at a certain early period of its development for the results it may in the course of time properly expect to attain. One of its duties is to aspire to show how the mind of the average man appears to him as it does and why.

In the same way the science of music has first to dig down to its foundations and show their form and connexions. Only then can it build upwards from these and aspire to give a full and true account of music as it appears to the musical mind, i.e. to the mind that is not crowded with scientific knowledge concerning music and actually thinking of it, but to the mind—even though it be the same mind or person—that is for the moment hearing and enjoying music in the ordinary way.

The science of music has for centuries paid the greatest attention to harmonics or upper partial tones. It has tried to explain many things by them. The insufficiency of the results has turned the hopes of theorists in later years towards the lower tones that appear in chords, namely to the difference-tones. But of all these things the ordinary musical mind is quite regardless in hearing and enjoying music. It is only with difficulty and effort that it can be brought to recognise their existence even when the attention is not aesthetically engaged. And when it is again so occupied, harmonics and difference-tones disappear from view entirely. That fact alone suggests the view that harmonics and difference-tones have not the central importance for musical theory that has often been claimed for them. And it confirms a theory that can find other foundations of greater validity.

This does not, however, mean, as some have seemed to think, that the scientific attention creates harmonics, or that harmonics come and go according to the inclination of observation. I say 'seemed to think,' for no one can venture to maintain such a view outright. We may want for our satisfaction to think that "the self sets itself" first and then all the rest of the world, including harmonics, according to its inclinations of self-realisation. If it is possible to leave this marvellous power to a Universal Self, we may well do so. But for our own self we must refuse to believe that it is able by mere change of attention to set anything into being at all. If harmonics are there when we attend to them, then they are also there when we do not attend to them. What we have to explain is why, when we attend to them, they appear in a different way than they otherwise do.

And that is not a difficult task. When we attend to a harmonic, we concentrate our inward gaze upon it alone to the exclusion of any setting or circumstances it may stand in. So we notice its own particular pitch and we can form a fairly sufficient estimate of its own particular volume. We may arrange for the independent production of a tone very like it and, by noticing the beating of the latter with the harmonic, estimate its pitch precisely. But we do not create it by our attention. For we have no knowledge from our will alone what its properties will be and real tones will not beat with fancied ones.

When we cease to attend specially to harmonics they are in themselves quite unaltered thereby. But if we then attend to the tone that contains them, we hear them in their full setting; we hear them as a part of the tone or chord we are attending to. And then, as everyone knows, they appear to us as the particular blend (or timbre) of the tone.

They do so because, being higher than the primary tone which gives the whole tone its musical pitch, they all fall within its volume. And in good musical tones, the upper partials are of a restricted intensity, wherefore they do not stand out prominently in the volume of the whole tone so as to call the attention specially to themselves. They leave the balance and symmetry of the fundamental still obvious. These qualities are no longer so perfect, of course, as are those of the pure tone. But they are far from being so vague and deteriorated that the sound could be mistaken for noise, in which balance and symmetry have been lost or at least made very hard to find.

The harmonics of a good musical tone only make a slight change of surface, as it were, in the whole tone. It no longer remains perfectly smooth like the pure tone; its volume acquires a character whose nature depends upon the harmonics present. A set of very high harmonics will give the tone a touch of highness or brightness. The lower harmonics will give more variety to the central body of the tone; it will not be empty and poor, like the pure tone, but full of interest and rich. If only the uneven numbered partials occur, the tone will take on another character, one that appears in the sounds produced by the nasal voice and by hollow cavities of various kinds, so that we associate the idea of hollowness with it, and call it a hollow sound. And so on.

The interests of music are not commonly served by sounds in which partials attract the attention to themselves or are separately distinguishable with ease. Those tones are the most valuable in which the minimal reduction of smoothness is compensated by a maximal richness and interest of pitch-blend. Tone must be rich and strong without being rough, and smooth without being dull or poor. It should be at once as full and as rich as may be. In analytical terms, the fundamental must be present in good strength to give the tone a fullness of the volume it 'aspires' to or is meant to be; and a typical series of partials should 'colour' it or give it a characteristic surface without being so strong as either singly or collectively to outweigh the fundamental or to stand forth in it so much that they take upon themselves the rank of primary sounds—tones actually played separately by the performer and written by the composer or intended to be heard separately.

It is unnecessary to recall that great variety of beautiful tone-surface is of the highest importance in music. These blends give a new interest to repetition and a new line of variation by which the hearer's mind may be led to give ear to the secret of the soul's life that the artist strives to convey.

The musical attitude towards harmonics, then, is the synthetic attitude. They create beauty when their synthesis is easy or inevitable, i.e. when their strength is so graded and unobtrusive that they appear to the attention only as a minor modification of the tonal volumes that compel the attention. But when the attention is used analytically— as when we pass from stone to stone of an architectural surface or from stroke to stroke of the brush in a picture—harmonics can be inspected singly. The rest of the tonal mass tends to disintegrate. The attention is then concentrated on the part and is scattered in the rest; whereas in a synthetic unity such as an artistic object the centre of attention is so placed that it radiates easily to the parts and binds them together in itself, while they point towards it and so make it easy to find rapidly.

Let us now consider difference-tones in the same relations. These lower partials, as it were, do not accompany single tones, so that they cannot play the same part in giving blend or surface to a tone as upper partials do. They appear only when at least two primary tones are sounded. Of course, they must also be produced by the interaction of upper partials as primaries. But the artistic subjection of these to the fundamental partial does not allow of their usually appearing in a single blended tone in any noticeable degree. Difference-tones are in any case weaker than their primaries, so that if they originate from the partials of a blend, they will be weaker than these and will therefore have less chance of being separately noticed in a blended tone than have its partials. Besides, the partials of the harmonic series could never produce either a difference-tone that was lower than the fundamental of that series or that did not coincide with some member of the theoretical harmonic series. So any new component of a sound that was produced as the difference-tones of its partials would only appear to be another partial of that sound.

The difference-tones that are due to primary tones are of quite considerable strength. This is true at least for the first difference-tone (higher rate of vibration minus the lower, or $h - l$) and for the second difference-tone ($2l - h$). The other difference-tones are much weaker and very difficult to hear, so that their inaudibility to the ordinary ear under usual circumstances hardly forms a problem for any possible theory of sound. In spite of their loudness these first two difference-tones are much less easily noticed than partials—as the much later discovery of them shows. Reasons for this obscurity are not far to seek. The strongest is the nature of their origin. They appear only when two sounds are played together. In order to detect them one has to consider carefully what exactly each of these two sounds separately contains and then to subtract the sum from what is heard when both are played together. Most persons do not trouble to do this. They expect the tonal mass of the two sounds to contain more than that of either, as it obviously does. But the peculiar overlapping and blending of simultaneous tones prevents them in most cases from discerning precisely the exact contribution of each of the primary components. Thus the new whole passes as the peculiar mixture of the primaries. This tendency is encouraged by the fact that in the octave where the fusion of the primaries gives the simplest product, there is only one difference-tone which is identical with the lower primary. So then where one might most readily have detected an addition, there is nothing new to find. And in the fifth, the only difference-tone is exactly an octave below the lower primary and so fuses with it as much as any tone could, thus making detection again difficult. A third reason lies in the fact that difference-tones are not under ordinary circumstances found to exist outside the ear, so that they could not be discovered, as were so many of the chief facts of acoustics, from a study of the movements and resonance of the sonorous body. They had to be found purely by the inspection of sound itself. And the attention was naturally directed in that to the primary sounds that were intended and played and upon which musical structure primarily rests.

Nevertheless difference-tones did not, of course, first come into being at their discovery. They were there all along, moulding the character of chords as a whole, giving them especially a touch of a largeness of volume that their primaries did not contain. The listener, as is often said, is unconsciously affected by them. That does not mean, to be sure, that his body or brain is affected by them, but not his mind. It means only that while the lowness is there in his sensations in a particular form, and he hears it as a lowness appertaining to the whole sound, yet he does not separate it out in its discrete form in the whole and *know* it as such. It would, therefore, be better to say that the listener is unwittingly affected by the difference-tones. For if sensation is a form of consciousness, and the difference-tones are in sensation, the hearer is of course conscious of them. If sensations are considered to be objects presented to the mind, which is only conscious of them when it knows them individually, then the hearer is *unconscious* of difference-tones even when he attends to the sound complex in which they appear before him, until he has separated them out from the group of sensory objects presented to him and has cognised them individually.

Having thus surveyed the outlying components of a sound mass, we may now deal with the relative importance of synthesis and analysis amongst the primaries.

It is to be noted first, however, that the objection brought against harmonics and difference-tones as the foundations and regulators of musical structure does not hold for our interpretations of the primaries. That objection is that the composer and the hearer commonly know nothing about harmonics and difference-tones and care still less, least of all when they are actually in the aesthetic attitude. They neither know of these things nor do they attend to them. But, while they have certainly not *known* about the volumes and coincidences and proportions of tones either, they have indeed always attended to them. For tone and interval are not *derived* from volume and proportion as from things that lie in a land beyond their own : on the contrary, they *are* volume and proportion. Whoever attends to tone and to interval, attends to volume in its balance and symmetry and to proportion of volumes. What our theory has done is neither to trace the heredity of tone and of interval and of fusion, nor to say what remote stars have influenced their horoscope; but it has dissected the very body of these things, as it were, showing what they consist of and how they are related to one another and to other similar things. Thus the change for the artist and hearer is merely from the practical and aesthetic attitude to the cognitive attitude—towards one and the same material. To skill and feeling is added knowledge. From using merely nominative terms for the objects of sense we pass to systematic terms, which not merely point them out on the basis of mental association, but which indicate their place amongst other objects and their relations to them on the basis of systematic knowledge. No objection can be brought against this knowledge from the artistic or practical point of view, for it builds upon the same ground as they do. It only adds the fullness of knowledge to the sufficiency of sense. Then sense not only is present with the mind and affects it to feeling and emotion, but it is known as well.

The primary tones of a chord blend with one another or with their fundamental in the same general way as do harmonics. They are much more easily recognised in the whole partly because they are louder, partly because they are known and intended to be played. They are part of the player's conscious intention, just as the blend or surface of tone in its synthetic form is. We have already seen how tones an octave apart may fuse so well together as to be mistaken for one. This high fusion of loud tones approximates to that generally valid for the weak tones of partials.

But the practised musician is able to pick out the primary tones of a chord with considerable ease. The most gifted ear can pick them out at once unfailingly in any part of the musical range and on any instrument. This analysis cannot, of course, annul the underlying synthesis of tones that is due to their volumic overlapping. But the gifted ear can at once seize upon the pitch predominances that the volumes contain, so as to cognise the component parts of the whole. A clear analytic view is obtained without any of the synthetic effects of overlapping or fusion being lost. Analysis, at its best—in dealing with primaries—does not require the finest ear to pass successively from one pitch-point to another in order to cognise them all. They are all grasped at once, as any of us grasps the whole of a simpler visual pattern or of a word in one gaze of fixation.

Nor is there here any general confusion between primaries and harmonics. For the latter are heard on all familiar instruments as synthetic blends, not as separate tones. If the grouping of tones makes one or other harmonic very loud, this will tend to be heard as a primary tone. But the prevailing attitude will be to distinguish only the loudest components as primaries and to hear harmonics in their usual blend. This attitude is greatly supported by the expectations made habitual by the general course of musical spelling and grammar, into which harmonics will not often fit coherently.

The less gifted ear does not distinguish the primary tones so readily. It may well learn to recognise each interval and chord as a whole, as a characteristic thing, as one learns to recognise words as a whole without reading, or thinking of, each letter separately. And it may also then readily learn to spell out the tones in the easier or more frequent groupings, so as to be able at least to name the relative pitches of each. But the first prevailing tendency is synthetic, even over and above the inevitable synthesis of fusion; analysis of (stationary) chords is, then, the result of effort and special attention.

But in melody the attention is almost relieved of any effort of analysis. The analysis takes place as a matter of course; or rather, the sequence of tones that we call melody is purposely so formed that the attention will follow it easily.

In primitive music there is commonly only one singing voice, which displays some distinct melodic form whereby its movements acquire unity and interest. In polyphonic music several voices proceed simultaneously, each one being melodically controlled in this way. In harmonic music the fullest melodic treatment is given commonly only to one voice, sometimes to two concurrently. The rest of the tonal mass of each moment is handled synthetically, so that the listener apprehends it rather as a whole. The sequence of tonal masses is regulated partly by the requirements of melodic form and partly by the relations connecting harmonic chords. Polyphonic music is, of course, also a sequence of (harmonic) chords; but the melodic treatment of all the voices leads to a predominance of the melodic connexions of the homologous voices of successive chords over the harmonic or fusional connexions within each chord.

The degree to which the harmonic and melodic aspects of music prevail over one another is thus very variable. At the one extreme we find each voice so perfectly finished melodically that both the artist and the auditor fail to apprehend the harmonic values of successive chords in any special way, although they in no wise fail to hear how far the different voices fit agreeably into one another's movements. The basis of this agreeable conjunction is, of course, the fusional relations of the tones of each chord. These are necessarily indestructible and irremovable by any treatment of the attention or by any abstraction. But nevertheless a special attitude of abstraction does lead the ear to make as little as possible of them for the production of the larger syntheses of the art. At the other extreme we find the melodic interest completely subordinated to the harmonic. Each chord is a fusional mass enjoyed for its special 'colour' or mass effect. The sequence of chords is not decided on the basis of melodic form in any specific sense. Melodic sequence, in general, is, of course, just as insuppressible and irremovable as is harmony. The various chords that follow one another must do so in such a way as to satisfy the minimal demands of melodic movement generally. They must move by as small steps as possible and must not cross one another, and so on. This minimum is enough to guide the ear easily from one chord to the next, but it is not enough to create any sort of melodic form. In fact the melodies that result may be perfectly irregular. They only provide enough obvious movement to guide the ear. Thus the mind is left free to devote itself to the harmonic interests of the music, and the artist or improviser may pay his greatest attention to building effects upon a synthesis of harmonic sequences.

The matter may be stated in a somewhat more figurative manner. Each mass of sounds that constitutes music in several parts or voices has two aspects : the one is its volumic aspect, the fusion characteristic of it as a whole, or in any of its parts, i.e. between any two of its voices; the other is its pitch aspect or its ordinal predominances, the points of sound that stand forth intensely in it. The art that builds up chords into complex music may, as it were, make either of these two aspects the surface of the product that is to be exposed to the hearer, while the other is made the mere surface of suture, cementing one brick of the building to another.

If the ground of artistic synthesis is pitch, great care must be taken in the selection of each brick that its pitch-points fit into those of the next, so that the sequence will give a perfect complex of melodic figures, easily surveyed by the listener. The subsidiary interests of the art require the sequent chords to be so harmoniously consistent that they will not severally fall to pieces or confuse the movements of the different voices and will yet knit together so as to make a stable whole. But the listener is not concerned with them beyond this.

If the ground of synthesis is harmony, that aspect of each chord will be turned outwards. Sequences will be selected specially for the manner in which they link together to form large harmonic surfaces or masses, as it were. The melodic aspect is required only in so far as it helps to bind the chords to one another on the unexposed surface and so to perfect the underlying stability of the structure.

Or, if you like, in the one style of music harmony is put in the focus of the listener's conscious mind, while melody remains in the background of it; in the other style conversely. Or again, in the one harmony is merely sensed and felt, while melody is built up into complex figures, inspected, and watched in all its changes, and consciously enjoyed; in the other the melody is merely sensed as an atmosphere, while the specific artistic structure is harmonic.

All this comes to the same thing as John Hullah's oft repeated dictum about the horizontal (melodic) and the perpendicular (harmonic) views of musical structure[1]. The figure of speech is here derived from

---

[1] v. 25, 106: "I use the word *harmony* as representing the successive results of an accumulation of parts. For of a *chord*, as an isolated fact, the old masters took little account. They were not harmonists at all, in our sense of the word, but *contrapuntists*; laying melody upon melody, according to certain laws, but uncognisant of, or indifferent to, the effects of their *combinations* as they successively came upon the ear. Their constructions were *horizontal*, not *perpendicular*. They built *in layers*, but their music differs from most of ours as a brick wall does from a colonnade," etc.

the structure of the printed music in which the component tones of a chord are written below or above one another, while melodies run from left to right of the page through the tonal points of each chord. The figure is, of course, not strictly applicable to what is heard. For the horizontal aspect is not spatial, as the adjective suggests, but temporal; the field of hearing—if the pitch series (or the length of tonal volume) be called its perpendicular dimension—has no horizontal aspect at all as far as music is concerned. But if allowance is made for this discrepancy, the figure is apposite—more so indeed than its originator could have known. For the harmonic dimension is really akin to a spatial dimension; it is ordinal, and space is probably an ordinal derivative.

We have thus characterised in general the relative importance of fusion and of analysis in music, and we have given these two aspects of, or attitudes towards, tonal masses a basis in the nature of these masses themselves as sounds. In other words we have shown upon what features of tones fusion rests and what points of tonal volume offer themselves for special analytic attention. In freely planned experiments these attitudes may be prescribed, or imposed upon oneself voluntarily and followed at leisure. As in other regions, so here it is found that some circumstances make synthetic apprehension easy, while others favour analysis or attention to a part rather than to the whole. The work of the musical artist is to bring these two attitudes under control, so that he may be able to guide the hearer's attention to any aspect of tone he pleases; or so to construct his tonal masses that listeners on the average will tend, with a minimal deviation, to devote their minds to those aspects of tone upon which the artistic effect has been built. For this purpose the artist must know as much as possible which factors favour each attitude and what power each factor has. Consequently the science of music is called upon to bring these factors into the fullest light of knowledge and to explain as exactly as may be how each one achieves its effect.

CHAPTER X

THE EQUIVALENCE OF OCTAVES

THE equivalence of octaves at first glance seems clearly to rest primarily upon the fact of the high degree of fusion appertaining to coincident tones an octave apart. For that is the ultimate fact of tonal hearing that most resembles the equivalence of octaves in music. Such pairs are very often mistaken for a single tone, more often than happens with any other interval. When octave tones are sounded in succession, there is of course no such approximation to the sound of a single tone, but there is an evident connexion between the two which reminds us of the transition from a thing to its replica, and which we therefore incline to call by the name of similarity or identity or equivalence or the like. Which of these terms is used, depends apparently upon the relative importance either for theoretical or for practical purposes that is ascribed to the sameness and to the difference of the two tones. For octave-tones are obviously not absolutely the same.

But although the primary basis of the equivalence seems so obvious, the system of facts of a similar nature does not seem to confirm it, at least in practical connexions. For the octave is only the highest degree of a series of grades of fusion which have been known more or less satisfactorily since the earliest days of the science of music. This series would lead us to expect a similar grading of equivalence, which by no manner of means can be claimed as real. We cannot call the tones of a fifth similar or equivalent as we call those of the octave, not even if we say the degree of similarity or equivalence is very much less than in the octave. It is true that crude and primitive forms of music do use parallels of fifths in the same way as we use parallels of octaves in our music. But even so the use is nothing like so extended, nor has it survived the first refinements of musical taste. Fifths are then no more equivalent than fourths or thirds or seconds are. But the equivalence of octaves is of the greatest and most extended importance in all music; far from being merely a primitive crudity, it increases in importance with the development of music.

Its central importance for musical practice and theory dates from the famous doctrine of Jean Philippe Rameau concerning the inversions of chords. In the preface to his simplification of Rameau's teaching D'Alembert pointed out that up till then work "had been confined almost completely to the collection of rules without reasons for them; there had been no discovery of analogy and of a common source; blind trial had been the sole compass of artists." "M. Rameau," he wrote, "is the first to begin to dispel this fog of chaos. In the resonance of the sonorous body he has found the most probable origin of harmony and of the pleasure it causes us : he has developed this principle and shown how the phenomena of music emerge from it : he has reduced all the chords to a small number of simple and fundamental chords, of which the others are only combinations and inversions; finally he has succeeded in apperceiving the mutual dependence of melody and harmony and in making it felt" (9, vi f.). "Whatever may be the fruit of the further efforts of others, the fame of the learned artist has nothing to fear; he will always have the merit of having been the first to make music a science worthy to occupy philosophers; of having simplified and facilitated its practice; of having taught musicians to carry into this region the torch of reasoning and of analogy" (9, xviii). Later on (9, 222) he wrote that a certain special difficulty and some others less considerable, would not prevent fundamental basses from being "the principle of harmony and of melody; as the system of gravitation is the principle of physical astronomy, although this system does not account for all the phenomena that are observed in the movement of the celestial bodies." The idea of the connexion between chords that involve the same notes of the octave is so familiar to the modern musical mind that it is necessary to recall clearly that the idea did not always stand in the forefront of the musician's cognitive consciousness. He may always have felt it, to be sure, but he certainly did not always know that he felt it[1] (cf. 60, 43).

---

[1] Readers who look upon the connexion of inversions as perfectly obvious will be interested in a quotation from a contemporary of Rameau's to whom the latter's doctrine came as a novelty:

"We must not omit an observation most easy to make at this point and also of the greatest moment for the clearness and solidity of the doctrine we have been gradually expounding. A concert has need for example of three voices if it is to embrace with their help three consonances, prime, third and fifth, which are called by the masters Harmonic Triad. The prime is always found placed in the lowest, the third in the middle, the fifth in the highest place. Now suppose that the prime is moved to the higher octave, so that the third remains in the lowest place. The ear is no longer satisfied with it. It no longer seems that the concert is finished. Hence the concord does not feel that it has yet returned to that note whence it has taken, and in which it recognises, its origin and in which alone it can come to rest and finish. It is openly apparent that the harmony is suspended.

Rameau himself actually thought that we fail to distinguish octaves in "the resonance of the sonorous body." The partials 1, 2, 4, 8 and 16 are, of course, octaves, which, he said, really resonate even more loudly than do those numbered 3 and 5, because of the size of the resonating parts of the musical instrument. So even though we fail to distinguish them, we are nevertheless necessarily affected by them, "but by an occult feeling that has so far prevented us from discovering its cause" (54, 3). From this feeling our sense of the identity of octaves has arisen. We actually prefer to have tones closer together than they are offered to us by nature in the series of partials, in order that we may have them within the range of the voice. For as we do not distinguish the octaves amongst the partials, the range of the voice is soon exceeded. Thus, 1, 3 and 5 take us up through a range of two octaves and a third, and they actually include only one interval less than the octave, namely the major sixth between 3 and 5. The ear also finds it easier to move about amongst close intervals because of the short distance between their tones. But the identity of octaves does not prevent them from intro-

This prime voice has nevertheless not been omitted. We have done naught but transfer it from the lowest place, where it stood, to the highest. We have still in ear the same three notes. How then has so great a change in the effect of all been made? To imagine hearing the low octave of a high note that we actually hear is very easy for anyone. So in this way we shall be able to make up for the defect, replacing in fantasy the true bass in the place whence it was taken. Then we shall have this principal voice present with us in two places: once in the high part where we hear it, and again in the low part where we imagine it. Our ear will nevertheless not yet be satisfied. We shall still not hear the perfect chord, the chord that concludes. And why so? Because the force of the high voice that is really heard prevails over the force of the low voice that would only be imagined. The sense of the ear that is the natural judge of harmony, does not let itself be deceived by the imagination. It would still refer the two prime voices that are actually heard to the third which is also sensed; and thus the harmony would still remain imperfect. It would not refer it to the imaginary fourth voice by reference to which alone the two higher notes could change their proportions and render themselves apt to conclude. This most simple observation which turns upon an experience known to everyone and beyond all doubt, proves that the common statement that the one octave is the equivalent of the other requires some limitation. In a large number of cases the statement is true, but not in all. In particular it is always false in reference to notes that do duty as bass; the which in changing place change their nature and make the nature change of all others from below which they withdraw.

Now if that is so (and nobody can deny it), however could truth or at least verisimilitude belong to the new doctrine of inversions which is nowadays so celebrated as a thing most useful to the art and perhaps the most noble secret that has yet been discovered in harmony?

To me there seems to be nothing to recognise in it but error and perversion" (57, 33 ff.).

It is clear from the above that the 'sameness' of octaves is not the same idea as the equivalence of inversions (cf. 60, 27).

ducing some differences into harmony and melody. But that, Rameau
said, consists "only in the different modifications of one and the same
whole differently combined, where sounds cannot change their order
without the help of their octaves" (54, 13). The sounding of an octave
in place of the fundamental in no way distracts the ear from the natural
whole that guides it; the ear recognises the fundamental sound in its
octaves, no matter what the order of the parts of the chord; it is always
reminded of this same whole (given in the resonance of the sonorous
body). If the chord is consonant, it is equally so in all its combinations.
In short, 2, 4, 8 and 1 are for us but one sound, in which 1 always
presides, whether we hear it or not (54, 16). Identity, Rameau added,
may seem rather an extreme term to use, but you may adopt any
term you like so long as, not going so far, it goes far enough.

That is precisely the difficulty in this problem—to find a theoretical
basis that will evidently go as far as is needful in establishing sameness
and with equal evidence refrain from obliterating the differences that
feeling and practice demonstrate. Rameau certainly overdid the aspect
of sameness. He admitted himself that we follow in our music the
traces given by nature in the resonance of the sonorous body "only
by the grace of these octaves" (41, 36). That is perfectly plain; it has
often been laid as a primary difficulty against those who claim to derive
the tones of the scales from the series of partials. How are you going
to bring them down from their heights to within the range of an octave?
Some second principle is obviously required for this purpose. This
was given for Rameau in his "occult feeling." Without that the needs
of the voice would remain unsatisfied; or rather the voice would have
had to be devised so as to cover a much larger range. And the ear
would likewise have had no scope for preferences as to the sizes of
intervals.

Rameau was also right in claiming that in the different inversions
we are reminded of a certain whole, but that whole is not the octaves
1, 2, 4, 8 and 16. Such an answer would be elicited from the mind of
no musician unlearned in the claims of theory. But any musician would
answer—more or less so—that the whole recalled to his mind is the
group of all possible combinations of the chord. When asked to name
a given one of them, he will say : "that is (the—arrangement of) the—
inversion of the—chord." When one of these combinations is heard,
the common relations are also 'in some way' heard, but the differences
peculiar to it are equally evident. Though the parts out of which the

chord is composed and 'in some manner' the whole that it forms are always the same, yet the consonance is by no means the same, nor is the harmonic treatment, although it may well be the case that inversion will not turn any consonance into a dissonance, or conversely. For the purposes of science, it is just the 'some way' and 'some manner' that is the problem. That is what we must give precise form to, so that the practical and sensory consequences so clear to the musician may follow evidently from as clear a conceptual foundation.

Helmholtz's theory has seemed to many to be a great improvement upon Rameau's or even to have finally solved the problem. In the octave we hear again a part—at best the half—of what we heard before—the fundamental and its own special series of partials. A most suggestive and winning explanation, very hard to abandon even when it has been disproved on other grounds! A theory is always seductive when it has a fair and clear speech for every phase of the business, for every doubt and hesitation, and withal so cleverly conceals the fact that the basis of explanation has only been assumed; this basis is not really patent and clear, as it is in the analogous cases referred to for support—the synthetic similarity of faces; it is only 'just as good.' What Helmholtz failed to show was why partial tones ever come to form a fused synthetic whole. And it is difficult for most folks to appreciate the importance of this omission. The explanations which flow from the assumptions are for all ordinary cases apparently so neat and apt that more could hardly be desired. Further demands and criticism look like finical pedantry.

And yet Helmholtz secured his whole basis of explanation by mere analogy—one of the kind that can be stated so plausibly for either of two opposite ends. For if in the octave we hear again a part of what we heard before, that should lead us in the course of time to distinguish the first and second partials of a tone as different primary tones. The progress of musical practice would thus bring about a gradual analysis of timbre into its ultimate constituents. Some psychologists believe that the world begins for the child in William James's words, as "one great blooming buzzing confusion" (27, 488). But we know that it soon clears up into its many distinct parts. Why should it not be so with the parts of musical tones or of ordinary tones? Separation and separate handling should here also lead to mental distinction and abstraction. Who is to hold the balance between the tendency to confuse the parts of a whole with one another or with the whole and the tendency to distinguish the separable parts of the whole from one another or from the whole? Helmholtz, after all, gets no further than does Rameau with this "occult feeling." In fact this phrase better conforms to the results of Stumpf's criticism of the theories of consonance given by Helmholtz and others and to the suggestions finally made by Stumpf as to the probable basis of fusion in some synergy of the nervous system. The "occult feeling that has so far prevented us from discovering its cause" is just what we might expect from 'synergy,' which is an occult (cerebral) process that has so far prevented us from formulating its nature.

In recent years an attempt has been made to account for the equivalence of octaves by setting up the series of differences that lie within the range of an octave, no matter what its general pitch may be, as the primary qualities of hearing. Then the series from c to c', whether it be taken continuously or discretely as in any specific scale, is a series of qualities like that of the spectral colours; only in the tones we do not merely just return to the starting-point but we are able to repeat the series a number of times.

This theory can hardly be discussed without close study of the psychological notion of quality. That includes all kinds of sensation such as touch, cold, warmth, pain, sweet, sour, the various smells, the colours such as blue, red, etc., muscular feeling, hunger, thirst, etc., etc., all as specific feelings, without concern for their intensity or localisation or for any other distinguishable aspects of them, except merely their kind. This bare kind or quality is the thing of all things that we know perhaps least of in itself. We seem to have some understanding of it in vision; but our understanding is here almost solely physiological. None of us knows what inner connexion, if any, there is between blue and red, or between yellow and blue, as felt colours. There seems to be none, and yet we at once recognise them all as colours.

One of the characteristic features of colours is their changes of kind. Red passes through orange, that resembles it, to yellow, that is like orange, but not at all like red. A similar change brings us to green, then to blue, and finally back to red through purple. If we are to consider the differences included within the octave as qualitative, comparison with colour would incline us to look for characteristic turning points, as it were, within the octave. These might be supposed to occur at the thirds, fourth, fifth and sixths perhaps. Various suggestions have been made. But none of them really explains the peculiarities that would be thus described[1]. It is, of course, conceivable that from a detailed study of classifications made in relation to the various forms of tone-deafness, etc., a good and probable physiological theory of tonal quality might in time be obtained, just as has been done in vision. Such a theory might then explain the peculiar relations that characterise thirds, fourth, fifth, and the rest. No very satisfactory explanation has as yet, however, been given even of colour affinities. The prospects of raising a lucid theory of music on this qualitative basis are, to say the least, not yet exciting.

Of course, that would be of no consequence at all if the classification as quality were logically inevitable. It is not so by any means. Some objections may be raised to the theory from the special difficulties it creates, from the obscurity of the ground it rests upon, and from the special phenomena of quality which the classification must introduce (cf. 77, 44 ff.). But until exclusion makes one theory or another logically inevitable, the merits of theories rest upon their respective powers of accounting for all the facts. The theory of octave qualities does not reduce the amount to be explained. The assumptions it makes require as much explanation and justification as do the facts they are supposed to explain. And a better explanation can be given without them.

The intervals in common use and their inversions are reducible to six pairs:

$$2, \text{II},\!-\!\!-\!3, \text{III},\!-\!\!-\!4,\!-\!\!-\!\text{T}$$
$$\text{VII}, 7,\!-\!\!-\!\text{VI}, 6,\!-\!\!-\!5,\!-\!\!-\!\text{T}.$$

The only marked change in grade of consonance produced by inversion is found in 4—5—4. The fifth is clearly more consonant or fused than the fourth. But in all but the tritone a decided difference is wrought in the interval itself. In the volumic theory of tone already developed

---

[1] As the octave according to the volumic theory is the greatest approximation towards the balance of a single tone that two simultaneous tones can make, and the fifth is the next, we might expect a certain parallelism in the character of the steps by which we pass from the two ends of the octave to the fifth:

$$o, \text{ VII, 7, VI, 6, } 5 \Big\rangle \text{Tritone}$$
$$p, \text{ 2, II, 3, III, 4}$$

Here the fourth is taken as the counterpart of the fifth, as it were. Otherwise the parallel will only hold if the fourth is slumped with the thirds and the tritone is set over against the minor sixth when it functions (in equal temperament) as a discord (augmented fifth). Thus:

$$o, \text{ VII, 7, } \quad \text{VI, 6, } 6^5, 5$$
$$p, \text{ 2, II, 3, III, 4, T, 5.}$$

Cf. Chap. XXI below.

we have good ground for the understanding of both these facts. The balance and symmetry of very different intervals may be approximately equal. But the intervals themselves are so different because they are quite different proportions of volumes.

Now the musical ear is not restricted to a knowledge of the simplest intervals. These are naturally of great importance in music, because they are amongst the simplest complexes of form known to the art. The simplest of all is the absolutely pure tone. A variant upon this is the blend, which gives the tone a surface, as it were. Interval is the first step that involves a variable proportion, constant only for each specific interval. But it is only the first step on a long line of possible complications, each of which may equally well be learnt as a definite complex of proportions, or as a 'pattern.' Let us follow out this process of complication, beginning with the addition to a simple interval of the octave of its lower tone. The 'chord' c, e, c^1, for example, may be represented thus (Fig. 3) :

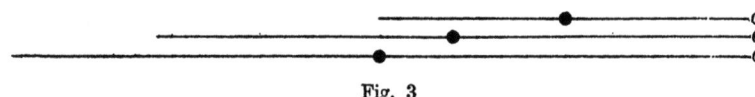

Fig. 3

The length of the lines represents the relative length of the volume of each tone and the middle points mark their pitches—the points that predominate in each volume and so give the whole a definite mark by which it can be placed in the series of all the tones. The volume of a tone is, of course, not homogeneous throughout its length, as the simple line suggests. It probably varies from the central pitch-point towards either end by a regular decrease of intensity. In any case this variation is quite regular, so that each tone may be a symmetrical whole. The range of the variation—from the pitch maximum to the opposite ends of the volume's length—will probably be the greater, the louder the tone is. Consequently when several tones overlap to form a chord, the intensity at each part of the chord's volume will vary infinitely according to the relative strength of the component tones. For the overlapping will give some sort of summation of the strength of each particle of sound that is common to two or more of the component tones of the chord. We cannot yet say precisely what the mode of this summation is. However, there is a feature of every chord that is in no way affected thereby, namely the relative or proportional position in the whole volume of the points where a departure from the regular changes that constitute the balance and regularity of a single tone occurs. And these points are bound into definite sets by their dependence on the physical stimulus of sound. They are always the same for any one ratio of vibration. And they are psychically fixed by the ordinal character of the points themselves. Thus the whole volume will be marked out into a set of proportional parts properly indicated in the diagram given above.

The musician in the course of his practice is made thoroughly familiar with the complex cec^1 both as a whole and in its parts, c, e and c^1, and their binary combinations, ce, ec^1, cc^1. In time he becomes able to survey these parts within the whole and to recognise their presence, either in an absolute way by naming their exact pitches, or in a relative proportional way by recognising the intervals they form. Even though he cannot banish c from the whole complex, he can survey the 'upper' parts around the pitch-points of c^1 and e and recognise the proportions of these. Or he may think of, and attend to, cc^1 and recognise its presence, ignoring e the while. Or he may dwell upon ce and ignore c^1 or at least its predominant parts about its pitch-point. This process of abstraction is already familiar in all those other senses which show a definite field or an ordinal system, such as touch and vision. We can shift the attention easily from finger to finger so long as touch sensations appear in either. In vision we are much more expert at such spatial or ordinal abstraction. The patterns of wall-paper often allow of combination and recombination in the most varied way.

The special peculiarity of hearing in this respect is the relative slowness with which the average person acquires practice and skill in recognising tonal proportions and in abstracting them from complexes of tones. In the visual field we can move patterns from place to place or rotate them, and dissect them as we please. In hearing rotation is impossible; movement within the sensory field is only possible in so far as pitch and volume of tone are altered in the way laid down by the physical stimulus; and dissection is limited in the same way. These restrictions make analysis so hard that most people are discouraged by them. But they are easily enough overcome by those in whom a good ear has created special interest and enthusiasm.

As to the way in which we may judge of the fusion of parts in the whole by abstraction from the whole, there is some difference of opinion. It has been urged that fusion—the degree to which a tonal mass appears to resemble the unity of a single tone—must necessarily be, and is, modified by the addition of a third tone to any pair. The mode of alteration will depend on whether the new tone forms a greater or less fusion with either of the two tones than they form with one another. Yet one might have expected the united fusion always to be worse, since the new tone necessarily forms a lesser degree of unity with either of the first two than that one formed by itself, being a tone, i.e. an optimal unity. Therefore when this deteriorated tone is added to the other one of the pair first given, the triad resulting should always be less fused than the original pair. Probably a good deal depends upon the point of view. If we look for mere plurality of tones, any trio will be more plural than the duo. If we look for the amount of good balance or fusion, as that is known in the octave, fifth, etc., we shall find more of it present in cgb than in cb. Here mere interpenetration over the whole tonal mass is not so much the standard, as perhaps a certain kind of interpenetration already familiar in various forms. There may be some abstraction in the process—i.e., a local abstraction within the ordinal field of sound of the chords. It is evidently not easy for those who experiment upon the fusion of more than two tones to make their point of view in observation quite clear. The opposite view has been upheld—that the addition of further tones makes no difference whatever to a fusion already given. This seems quite a reasonable position provided the above-mentioned 'local' abstraction of fusion has become easy enough.

There seems to be no reason in the nature of tonal complexes themselves why such abstraction should not succeed with those who are highly gifted and practised acoustically. They would then isolate for attention the tones in question and see the sort of balance and symmetry they possess. Of course they cannot lift the tones they abstract out of the whole complex they are abstracted from. Abstraction here means only devoting special attention to certain tones and recognising in them features that are usually characteristic of them in isolation, in so far as these features have been only partially or not essentially distorted by the presence of the other tones. Thus one who abstracts cc^1 from the chord cec^1 will notice that the maximal volume of the chord is that of c; that c^1 is present as a pitch at its usual ordinal place, that the parts of the tone lying around the pitch-point have the proper tonal symmetry and that there is the usual clean function at the pitch-point of c. Of course the tone e will often be encountered during this process. But one who is highly practised may pass to and fro about this irrelevant tone without being disconcerted by it, and may feel as able to give his judgment as he would if it were not there. For others, however, the third tone may be a source of great disturbance and they may feel they never really can ignore it, so that it always spoils the effects for them.

Later on we shall meet evidence that will call for a more special effort to settle the question of the part played by the fusion of single intervals in chords.

The 'chord' cec^1 is in the experience of the musician not only given at all levels of pitch, but when it occurs at any pitch, it then commonly occurs at the octaves above and below. Thus cec^1 may be carried up and down over the piano, $cec^1e^1c^2e^2c^3$ (Fig. 4). The new parts here make no significant change in the diagram of volumes. e^1 fits in between the pitch of e and the common upper limiting point of all tones, while c^2

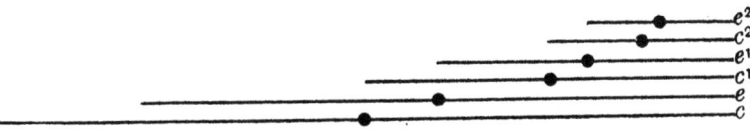

Fig. 4. Illustrating the continuity of 'pattern' made possible by the volumic relations of the octave and upon which the connexions of inversions of the 'same' chord rest.

likewise fits in between the pitch of c^1 and that point. Both of these latter tones can be taken as mere appendages of their lower octaves, the more so the weaker their strength is. In any case the new tones do not spoil the previous pattern, but merely continue it further towards the upper limit of hearing in the same characteristically proportionate form. The component tones of the whole are not more easily separable for their being so many but less so. Only, the characteristic pattern of the whole remains the same and can be recognised with almost, if not quite, equal ease.

Now if ec^1e^1 or its extensions are given in the same way, they may not only be analysed into the same musical components c and e, but they give a pattern which is partly the same as that of cec^1. The closeness of the resemblance is the greater, of course, the further up the pattern is extended, $ec^1e^1c^2e^2$, etc. But it is obvious that cec^1 and ec^1e^1 are identical in respect of their common part ec^1. This part will make them thus far similar. And the resemblance is increased by the similar way in which the other tone is related to one of the tones of the common pair. Of course this sort of similarity is evident both in the mere musical symbols and their arrangement and in the common musical consciousness of our time. Our concern here is to show definitely how this similarity is grounded upon the sensory material of hearing, in the tones themselves. The similarity that appears to the musical mind is therefore not merely the result of musical analysis or of theory or of thought, but is a true representation of the relations of the parts of the sensory stuff of music.

Thus we do right to consider cec^1 and ec^1e^1 in a certain respect as mere aspects of one another, or to consider ec^1e^1 as a trifling alteration of cec^1, which for some reason we look upon as the normal or more fundamental form. The same holds for any other chord, no matter how complex and discordant. ceg in a certain respect appears again in egc^1 and in gc^1e^1. They are all patterns that may be said to be parts of their common extension $cegc^1e^1g^1c^2$, etc., except that the common pattern does not begin at the same part of its cycle, so to speak.

The musical consciousness that has got thus far, will find it easy to see the same pattern even when its parts are scattered more widely through the octaves of its extension. Thus we come to forms such as cge^1, ce^1g^2, ge^1c^2, the familiar positions of the various inversions. The connexion of these with the fundamental pattern ceg cannot remain obscure after the musical mind has learned so much as to be able to create ceg itself, to know it and to use it. Of course this pattern is only relatively feebly indicated in cg^1e^2; but for the practised musical mind— and we are here dealing only with practice in the simplest things, though for the foundations of the science they are the hardest problems— the connexion is as plain as daylight, as plain as is the ordinary hand-writing of our own language in spite of its great variations from the copperplate model.

In ordinary music, moreover, there is a much greater resemblance between cg^1e^2 and ceg than appears in the great intervals between the parts of the former, or in the diagrammatic representation of it on the basis of absolutely pure component tones. For these fundamental tones are ordinarily accompanied by upper partials. If we suppose merely that the lower partials are present in some strength, we should get

	c	c^1	g^1	c^2	e^2	g^2	$(b^2\flat)$	c^3	(d^3)	e^3	g^3				
c															
g^1			g^1			g^2			(d^2)		g^3	(b^3) (d^4)		g^4	
e^2					e^2					e^3		(b^3)	e^4		$g^4\sharp$
Total	c	c^1	g^1	c^2	e^2	g^2		c^3		e^3	g^3		e^4	g^4	

In this series the pattern ceg is represented more than twice. The resemblance would therefore, be more evident in instrumental tones than in pure tones. The musician here is not expected to analyse partials—a thing he rarely does at all. The point is only that these partials will reinforce an effect that is apparent enough to him already on common psychological grounds without partials. Partials alone would not suffice, without the volumic basis. But granted that, they will only repeat and confirm it. The same holds true to some extent for the difference-tones. Thus between tones whose vibrations stand in the ratio of $1 : 3$ (e.g. c and g^1), the first difference-tone ($h–l$) will have the ratio 2, and so will form—even in the case of pure tones—a link towards the filling out of a pattern and the extension of connexions by fusion beyond the octave.

But in all this we must not omit to notice that there are marked differences between the different inversions and their different positions. Therefore we observed above that these forms were identical 'in a certain respect.' Apart from that and for other purposes, their differences are great; and naturally too. The bass is, as above explained, the weightiest part of the chord, its centre of gravity so to speak; and it must make a great difference which part of the basal pattern bears this function. That is perfectly familiar in musical practice and theory, and just as clear on our theory. The pattern set by gc^1e^1 continues as $g^1c^2e^2$, etc. It has the same series as ceg, once it is well started; but, as given, it designates the gc^1e^1 complex unit of pattern most strongly. If we care to ignore that designation or are specially led to do so, then we may well see the ceg type most of all. What the special differences between these types of the same basal pattern, as it were, consist in essentially, we shall endeavour to show as we proceed.

We have thus shown that the equivalence of octaves rests upon a sufficient natural basis in the stuff of tones themselves. And our account in no way inclines us to underrate the differences that exist in that sensory stuff between the different groupings of tones that are equivalent. There is only equivalence for certain purposes. A point of view, an attitude, or a certain trend of abstraction has to be made for the equivalence to emerge so. strongly as to suggest sameness. Other attitudes may concentrate in other ways, and see practically nothing but difference. For certain purposes there is familiarly a very considerable difference between ceg or egc^1 and gc^1e^1.

The equivalence thus established for octaves does not apply to any other interval in the same way. Suppose we double the interval of the

fifth—cgd^1 (Fig. 5). The second fifth does not fit into the first so as
to be a mere repetition of its pattern. It would, no doubt, do so, if
each tone consisted only of the half of its actual volume that lies on
the upper side of its pitch (u in the diagram and not l). But these l
parts break into one another irregularly. The l-end of the second fifth
strikes in between the pitch-points of the two lower fifths. In the case
of the octave the l parts of the higher tones merely repeat or emphasise
the pitch-points of their lower octaves, so that no new or disturbing
element is introduced. The higher tones merely carry onwards and
upwards the pattern already given by the simple interval or chord.
Of course we can attend to the one or other fifth in the whole and
hear it as a fusion, but the two do not follow upon, or fit into, one
another as a continuation of one whole pattern. If gd^1 is played after
cg, and the attention is concerned with their justness as fifths, gd^1
will be heard as the repetition of cg. Or if the attention is concerned
with the indirect relation holding between c and d^1 through a real or
imaginary g, it will take a similar attitude. But if gd^1 follows cg as a

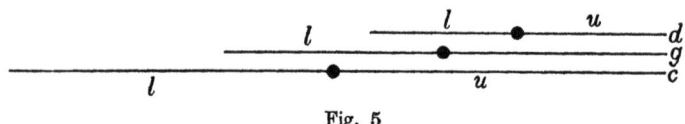

Fig. 5

part of a whole to which both belong, the attention directed to this
pattern will not find itself rewarded. This shows more clearly than
does the octave that equivalence of octaves is not the mere repetition
of a tone of the same 'quality' absolutely inherent in itself, or of the
same single interval of two tones; but it is the presence of a pattern
of which a chord of any number of tones forms a part, or the indication
of that pattern in a way that suffices for the musical ear under the
circumstances of the moment.

As the fifth is thus distinguished from the octave, so are the other
intervals. The octave is the only interval that thus makes possible
the extension of patterns. And it does so because it packs the repetition
entirely into the upper half of its lower tone, making only one new point
of predominance at the upper pitch-point. That is why the octave
is of such fundamental regulative importance in all music, and why
the equivalence of the octave, instead of being a survival from primitive
forms of music, is of constantly increasing importance.

The explanation we have given also shows that the equivalence
of octaves does not rest entirely on their great fusion, as such. Of course

the reason for the great fusion is closely allied to that which makes equivalence possible. But the equivalence does not rest on the balance of proportions. For if it did, the fifth as already noticed, would provide an equivalence of second grade, which is not really found. The fifth holds a steady second place in music to the octave only as a fusion, i.e. as a consonance in our music, or as a form of homophony in primitive music. Equivalence rests, not upon the balance of the parts of a single interval, but upon the way in which octaves extend the pattern of proportions of an interval or of a chord without distorting that pattern. Equivalence thus introduces an important new form into musical structure. This form is present, indeed, *in nuce* in the single interval, but it only emerges clearly as an important specialty when chords are freely used, and when they have been for some time steadily apprehended for the purposes of musical structure in a special way, i.e. not as fusions, but in another way, which we have classified as 'pattern.' Thus we can now well understand why the notion of the equivalence of inversions did not take clear shape in the musical mind until the period of harmonic music had been fully inaugurated and had had time to ripen into conscious formulation in Rameau or his more immediate precursors.

In other words that make it almost a truism, a system of connexions like those of inversion is only possible when the terms connected have become familiar. Simple though this is, it is so important that it may be set up almost as a principle for the study of chords in so far as the notion of inversion reduces these to a manageable number. We can speak of inversion properly only when we know that the best (average) listeners are so familiar with the different chords as to be able to recognise them readily as parts of the same pattern. After all any set of notes whatever can by suitable (octaval) transposition be inverted into a series of major or minor thirds with appropriate omissions. No real system of chords can be founded on such merely formal considerations. The primary factual question for every system of chords that uses the notion of inversion is : are the inversions recognised by direct hearing as parts of one pattern? Contrariwise, the formal reduction of all chords to columns of thirds does nothing at all to prove either the real importance or the real primacy of the third in musical structure.

The preceding exposition cannot, of course, prejudice the efforts of abstraction in listening to chords that may still become possible to the musical mind. It is conceivable, for example, that a mind might contrive to attend to a column of simultaneous fifths or of any other interval apart from their mutual interference and blurring. Great concentration and practice in distinguishing pitches—rather than intervals—might lead this way. Those who have absolute ear often recognise intervals rather by inference from their absolute pitches than by direct apprehension of intervallic proportion. So a mind might come to hear chords as columns of pitches standing at proportionate distances from one another rather than as volumic patterns of proportional nature throughout. Of course much that is of the greatest value, if not essential, to music, would thereby be abandoned— all harmonic effect in particular. Perhaps some of the latest experiments in music-making tend in this direction; for example Scriabin's columns of 'fourths.' Only the further developments of this line of construction and the general judgment passed upon it in the course of time will show whether it has struck upon new and useful faculties of musical analysis that will serve the synthetic ends of artistic creation. If we cannot discover whether Scriabin had a special attitude of listening to his own music, we shall have to see whether in time such an attitude will not prove to be essential for the artistic apprehension of his works.

CHAPTER XI

CONSECUTIVE FIFTHS

THE rule forbidding consecutive fifths is one of the fundamental generalisations of musical structure. The view is indeed sometimes expressed that modern developments have swept all the rules of harmony away, and that this one like others no longer holds because composers break it repeatedly. It is true that the rule has its exceptions. But the special means required to make such exceptions tolerable and their late appearance in any frequency in the highly developed art show that the rule has really the fundamental importance commonly ascribed to it. The breaking of an established rule is naturally the first fact to engage the attention, when it has been broken. The next question inevitable for a mind that feels the good effect produced in spite of the breach of rule is : what other elements of the whole in which the fifths appear, are responsible for the good effect? The pleasantness of fifths in a certain setting by no means discredits their prohibition under most circumstances. This could be gainsaid only by the pedant who lives on rules and does not apprehend the structures he studies in their primary aspect—aesthetically—at all. But the sole standard of art is the beauty inherent in the created object. We do right to expect art to be, like nature, a realm of law and order, not the sport of chaotic chances; and the study of its laws is the science of art. Rules are merely the expressions of the probable sequences of cause and effect already recognised. They are useful because in many cases they foretell the effect with accuracy. But if their prophecy is false, they must be corrected by a further study of the new effects, under the assumption that the effect is not the result of the one cause stated in the rule, but is the resultant of a number of causes, some of which act in opposition to one another and so produce from time to time apparently contrary effects.

The prohibition of consecutive fifths appeared comparatively early in the history of the art, much earlier for example than the formulation of the connexions of inversions. In the music of the ancient Greeks, series of fifths or of fourths seem to have been freely allowed in instrumental, but not in vocal music. This was known as the 'antiphonic' style. In vocal music only octaves were run in series; no other consonance was 'magadised' (16, 21, 154 ff.). On the common instruments of Greek music, the lyre and the cither, sequences of fifths or fourths were only possible in so far as the *melos* lay within the lower tetrachord of the octave; for in their music the only distinctive melody lay *below* the accompaniment. After the highest available fourth had been reached, however, the accompaniment remained stationary in the highest tone of the instrument, while the melody wandered at will even into unison with it. When the melody again descended out of this region, it drew the accompaniment with it in fifths or fourths, only the final interval being always the octave (16, 232 ff.). According to Aristotle music involving different intervals ('symphonic' style) was less pleasant than the antiphonic.

Greek music thus seems to have been essentially monomelodic. Vocal melody was evidently absolutely single (cf. 16, 157). And although in the instrumental style a further approach was made to polyphony, especially when different intervals became obligatory in the upper tones, yet it was clear to the Greek ear that the *melos* still lay unobscured below the accompaniment. The latter did not itself form a voice (16, 234).

This state of the art forms a most interesting parallel to the earliest forms of Western music known as organum. In its strict form this consisted simply of series of fifths or of fourths, or of these primary voices doubled at the octave, the upper an octave below and the lower an octave above. The very difficulty that probably prevented the Greeks from magadising in fifths or fourths, namely the occurrence of a tritone instead of a fifth or a fourth once in the complete scale, may have been responsible for the development of a 'free' organum, in which the 'vox principalis' moved from unison with the 'vox organalis' up to the fourth while the latter remained stationary, and the like (81, 51 ff.). The variant thus attained was then preferred for its own sake and developed to greater freedom. And after a time the only other possible relation of voices—that of contrary motion—seems to have appeared quite suddenly (81, 71 ff.) :

The earliest known expositions of the new organum are contained in the *Musica* of Johannes Cotto, written about the year 1100.... The organum, we find, is now constructed entirely of consonances, and the arrangement of these is decided chiefly by the various kinds of progression adopted by the voices.... Although the similar [parallel] motion of the voices is by no means forbidden, a contrary progression is on the whole preferred (81, 77). (Hic facillimus ejus usus est, si motuum varietas diligenter consideretur: ut ubi in recta modulatione est elevatio, ibi in organica fiat depositio et e converso. 83, vol. 150, 1429.)

But the series of consecutive consonances of the same kind did not go beyond two or three. "Existing compositions prove that the first actual expansion of the polyphonic principle, the addition of a third part to the original two, dates from this period, and that the fourth part followed soon after" (81, 85; cf. 44, 79).

There cannot be the slightest doubt that music in three or four parts in which contrary motion prevails is polymelodic or polyphonic. There is not now any such difficulty in following the various voices as Plato[1] and Aristotle complained of in the Greek music of mixed intervals (16, 149 f.). And it is a noteworthy fact, which our further analysis will illuminate in a far-reaching way, that the decline of the antiphonic style and the gradual emergence of the prohibition of successive fifths and octaves, etc., proceeded in close relation to the development of distinctive polyphony.

Thus it appears that musical art can proceed only a little way before it comes to a distinct apprehension of the bad effect of consecutive fifths and before it makes their prohibition a primary principle of construction. We are not by any means, however, compelled to suppose that early Greek and Western musicians took perverse pleasure in ill-sounding experiments in symphonious singing. An isolated perfect consonance has at all times a beautiful aspect that reveals itself very readily to the mind, although comparison with some other intervals—when they have been found and fully appreciated—may make it seem thin and poor. But to the natural uncritical ear a high grade consonance is beautiful. If that beauty is made the object of great attention, it is possible that it might maintain itself for some time in forms of usage that at the same time presented latent aspects of ugliness. We have only to suppose that the latter had not yet caught the attention. Besides, this trend of attention would be prevented by another feature of consonances—the unity of voice or tune to which they in their grade approximate. The voices of men and women singing the same melody will fall into octaves because of the ease and unity thus established. The voices of women or of men, if they differ from one another in pitch considerably, would for the same reason tend to fall into the next greatest consonance—the fifth,—as the octave would not lie near enough to their average difference to attract their voices to itself; or if it did, one or both voices might be subjected to too much strain. Untrained

---

[1] Plato perhaps was not specially gifted musically (cf. 66, 17). But it does not require any exceptional musical faculty to follow two simultaneous melodies, if they have been properly composed and performed.

singers have been heard to sing in fifths. Stumpf recorded this of two maids at work in his domestic kitchen (52, 239). In primitive music also sequences of fifths have been variously established (*ibid.*).

An interesting experiment in such music has been recorded by Gevaert (1895, 15, 423) :

> Sequences of fifths, produced without thirds and performed slowly by very true voices, have nothing disagreeable about them. Consecutive fourths produce at first a bizarre effect, but the ear soon accustoms itself to that. I made a personal experiment with this on the 8th of July, 1871, at an archaeological gathering arranged by my friend Aug. Wagener, the eminent hellenist, in the ruins of the Abbey of St Bavon at Ghent. On this occasion I had a choir of men and children perform several diaphonic specimens of the two species [strict and free Organum]; the impression made on the audience, about a hundred persons, was profound. Everyone was unanimous in finding in this threadbare harmony a penetrating atmosphere of very remote antiquity. It is true that the place lent itself admirably to an evocation of this nature.

If we can thus show why sequences of fifths are for some time in the earliest stages of the art not only tolerable, but more or less inevitable, we must endeavour to find out on what basis the unpleasant effect rests that soon appears in the further development of the art. This problem has long been the object of inquiry and debate, and it is well that we should consider carefully what grounds of explanation have already been advanced.

The following are the chief theories of the prohibition :

1. *Habit and tradition.* This theory was advocated by W. Pole (50, 283 ff.; 17, 113 f.). A ready reason for the prohibition of consecutive octaves is found in the fact that counterpoint is a series of different melodies going together. Using sequences of octaves in counterpoint thus means professing to keep melodies different throughout and yet not doing so. But the rule as to fifths has always been a great puzzle, he says :

> It is asserted and generally believed that there is something naturally repugnant to the ear in such successions…. But still it is undeniable that any series of musical sounds will be accompanied naturally by consecutive fifths as well as by consecutive octaves; and with this example in nature before us, it certainly seems difficult to say that such sequences are forbidden by natural laws.
>
> We are bound to distrust here the appeal to the ear…. It cannot be denied that a succession of perfect fifths in counterpoint sounds very objectionable to musicians. But it must be recollected that from the first moment any musician began to study composition, he was taught to hold consecutive fifths in abhorrence; and it is to be expected that the result of this must be to make him believe that they are naturally

objectionable. If there is really any physical or physiological cause for the antipathy, it ought to be capable of being *shown*; if it cannot be shown, we have a right to presume it is merely the effect of education and habit.... We know one thing by experience, namely, that these fifths do not sound offensive to those who happen to be ignorant of the rule against them. There are many persons who have learnt music practically, and have been accustomed to it all their lives, but who have never had a lesson in harmony or composition; and if such people attempt to write music in parts, they will use consecutive fifths without the slightest hesitation, and not see anything objectionable in them;—rather a strong argument, it would seem, that the objection arises chiefly from a knowledge of the rule.

Even so notable a writer as F. A. Gevaert has given support to a kindred view in writing : "Influenced by the school rule that prohibits the succession of several perfect consonances of the same species, the musicologists have not failed to declare the diaphonies of the epoch of Hucbald and Guido as intolerable and monstrous. That is a counter-pointist's prejudice" (15, 423); and : "it is a modern prejudice to believe that sequences of consonant fifths as such jar on the ear" (16, 158). In support of this he tells of the experiment already quoted (p. 84, above), but he does not add any further justification of his view.

Pole's theory is, of course, very extreme and may be opposed on every count. As C. Stephens pointed out, the harmonic fifth is so prominent in certain cases, e.g. on stopped organ pipes, which give the alternate harmonics, that it may make that timbre unsuitable with music that would tend to direct attention to its presence, e.g. when a fugal subject is being given out in the lower part of the instrument (17, 115). G. A. Macfarren declared that the sequence of fifths is

repugnant to us at the present time, and not in this room alone, not in this country, but throughout all the civilised world wherever music is studied, and wherever it has resolved itself into a language instead of the barbarous jargon of savages. I cannot suppose that, as long as the organs of hearing have been the same, persons can have experienced pleasure many hundreds of years ago in progressions which are entirely offensive to us who hear them now; that the same acoustical properties, whatever they may be, which make them offensive in the nineteenth century could have been absent in the tenth century; and that progressions which through their as yet undiscovered properties are cacophonous to us can have been acceptable to the persons who heard them (17, 117).

Of course many persons may have "learnt music practically and may have been accustomed to it all their lives" and yet may never have attended to it analytically or aesthetically at all. Just as there are so many who are hardly even aware of the diurnal changes in the colour of objects in spite of their having attended to these objects in many critical practical ways which would seem to a colour artist to make such ignorance impossible. Think of the crudities such persons would produce in a first attempt at water-colouring!

But the weakest point of Pole's position is that, while demanding from the defenders of the prohibition an exposition of the physical or physiological cause for the antipathy, he omits himself entirely even to suggest the need for a cause of the convention, as he thinks it, by which in the first instance sequences of fifths came not only to be forbidden but also to be so heartily disliked. Such a convention to dislike requires as strong a cause as any unauthorised hatred, more especially as opposition to the convention does not seem, even in Pole himself, to have led to a change of taste for consecutive fifths.

2. *Excessive sweetness.* According to W. H. Cummings (17, 114) John of Dunstable, an Englishman, forbade consecutive fifths :

not because they are so objectionable, but because they are so sweet, so that the ancients could be really cloyed with the sweetness of the fifth. We know that fully to the end of the thirteenth century most of the harmony we can find consists of fifths and octaves. They found it so sweet that they thought it was time to leave it off. John of Dunstable is really the first who wrote against the use of them.

Or as G. A. Macfarren expressed it : "John of Dunstable said they were too beautiful, too much beauty could not be permitted, therefore, a succession of these delights was overpowering to the human senses" (17, 118).

Against this theory Sacchi wrote (57, 6) :

No one ever denied, nor shall I, that successive sweetness can change to displeasure. We can therefore well understand that a continued series of ten or fifteen fifths ought to displease and disgust us; and that would not be improbable. But that one single repetition of the fifth, merely by reason of its great charm, must suddenly displease and offend us, is not intelligible.

That is quite true. But we have not only to refute the theory, but to account for its formulation as well. And it is not difficult to see the motive of it. The fifth, as was noted above, is the second best consonance or fusion, and as such has a special beauty and sweetness, not necessarily under all circumstances of comparison, but at least under some. Of these circumstances probably only the latter group will control the judgment of the primitive ear and of any mind that is for the time being more or less uncritical or forgetful of the specially interesting intervals and chords that music has developed. In any case there is no doubt that the primitive ear is fixed upon the beauty of the great consonances, and naturally endeavours to make its art out of these elements. No doubt too for a time its attention to them prevails over any other features their sequences may create. But, of course, these features, being so pronounced, soon force themselves upon the attention. What is more natural, then, than that the only known feature of these intervals—their great consonance—should be taken as the ground of explanation, and that the theory should be : in sequences of fifths there is too much of the fifth's consonance? It is not that the ears of these early folks were undeveloped or different from ours, and that their minds were crude and lame; but their attention had been set into a certain direction by the course of the art till their time; and their minds naturally followed its suggestions. Looking backward is not nearly so difficult as groping forwards and accommodating soul and mind to new developments quickly. We must judge leniently when we think how long all our theories have tried to nourish their energies on the very poor diet of the harmonics.

3. *Want of variety.* Zarlino wrote in 1571 (82, Part iii, chap. 29, p. 216) :

The most ancient composers forbad the placing after one another of two perfect consonances of the same genus and species bounded in their extremes by one and the same proportion, while the modulations moved by one or more steps; as the placing of two or more unisons, or two or more octaves, or two or more fifths, and such like; ... for they well knew that harmony cannot spring but from things mutually diverse, discordant, and contrary, and not from such as in every way agree.

The composer, he says, must imitate the beauty of nature which makes no two things of any species exactly alike.

The explanation given by Helmholtz includes this one beside others :

The accompaniment of a lower part by a voice singing an octave higher, merely strengthens part of the compound tone of the lower voice, and hence where variety in the progression of parts is important, does not essentially differ from a unison. Now in this respect the nearest to an octave are the twelfth, and its lower octave the fifth. Hence, then, consecutive twelfths and consecutive fifths partake of the same imperfection as consecutive octaves (20, 359).

Only, the case is worse because the accompaniment cannot be carried out consistently without changing the key. (The *a* of the key of *c* is familiarly a little flatter than the just *a* of the key of *d*; thus two just fifths *c–g* and *d–a* would mean a departure from the scale of *c* in the *a*.) "Hence an accompaniment in fifths above, when it occurs isolated in the midst of a polyphonic piece, is not only open to the charge of monotony, but cannot consistently be carried out" (20, 360).

Helmholtz then proceeds to explain that :

when the fifths are introduced as merely mechanical constituents of the compound
tone, they are fully justified. So in mixture stops of the organ.... It would be quite
different if we collected independent parts, from each of which we should have to
expect an independent melodic progression in the tones of the scale. Such independent
parts cannot possibly move with the precision of a machine; they would soon betray
their independence by slight mistakes, and we should be led to subject them to the
laws of the scale, which, as we have seen, render a consistent accompaniment in
fifths impossible.

For the same reasons the second inversion of the major common chord
"expresses a single compound tone much more decidedly than 'the
first inversion,' which is often allowed to be continued through long
passages, when of course the nature of the thirds and fourths varies"
(20, 360). The second inversion may be represented as the third,
fourth, and fifth partials of a compound tone, the first inversion as
'only' the fifth, sixth, and eighth. But, as F. E. Gladstone pointed out
(17, 102), this argument will not apply to minor chords. The minor
chords have always been a thorn in the flesh of the harmonic derivations
of music.

The theory, then, argues that close approximation to the constitution of a tonal blend of fundamentals and partials makes sequences
of fifths and fourths admissible. These sequences are forbidden only
between distinct parts, because we expect independence and variety
from them, not monotony (cf. Gladstone, 17, 105). The theory suggests,
but does not state explicitly, that the prohibition of consecutives is
the stricter the nearer the interval in question lies to the fundamental
component of a blend. Thus the fourth is prohibited, "but with less
strictness" than the fifth. "Even thirds" have been forbidden as an
accompaniment (cf. *ibid.*). The theory is apparently consistent logically.

What is hard to understand is how the relation to partials creates
such unpleasantness in this case while in single consonances it makes
for harmony and pleasantness. Of course every theory must appear
in dealing with this problem to pull from the storehouse of explanation
contrary results for what seem very similar objects. That is a mere
restatement of the fact that isolated consonances are pleasant, while
sequences are often ugly. What every theory, however, must avoid
doing is using the same unaltered ground as an explanation of contrary
results. And that Helmholtz seems to do. If "hearing again a part of
what we heard before" is a ground of consonance, it is unlikely that
this alone would produce the ugliness of sequences of fifths. Monotony
is an idea hardly adequate to the effect to be explained. As Sacchi

might have said : we could understand that ten or fifteen fifths in sequence would have been boredom, but we should hardly take offence at two. Helmholtz seems to have felt this himself somewhat; for he inclines in part to Pole's view, saying :

> The prohibition of consecutive fifths was perhaps historically a reaction against the first imperfect attempts at polyphonic music, which were confined to an accompaniment in fourths or fifths, and then, like all reactions it was carried too far, in a barren mechanical period, till absolute purity from consecutive fifths became one of the principal characteristics of good musical composition. Modern harmonists agree in allowing that other beauties in the progression of parts are not to be rejected because they introduce consecutive fifths, although it is advisable to avoid them when there is no need to make such a sacrifice (20, 360).

Here he has not reached the point of view of some who claim that in these exceptional cases it is not a matter of admitting the ugliness for the sake of the beauty, but of outweighing the ugliness so that it no longer appears or even of creating positive beauty (cf. 35, 84).

The argument from want of variety is weak in so far as it has to meet its own objection for every interval; including sixths which by no manner of means can be claimed as forbidden in sequence (cf. C. Stephens, 17, 115, and G. A. Macfarren, 17, 119, who refer to both thirds and sixths). Sacchi added that

> it is true that on the false principle that two fifths displease for lack of variety, some [e.g. Zarlino[1]] have drawn the false conclusion that similarly the repetition of thirds and of sixths ought to displease when they are of the same species, and have therefore forbidden it; but their prohibition was not accepted by composers, who, disciplined by experience, carefully avoid repeating fifths, but are not in the least concerned about the repetition of thirds or sixths of the same species. Vain is therefore likewise the reason that is drawn from the desire for variety, which if true would be equally so in all the consonances; it would hold rather more in the imperfect than in the perfect; because after all it ought to be more tolerable to the ear to linger on the sweeter consonance than on the less sweet (57, 7 r.).

The validity of the argument, therefore, vanishes entirely in so far as it is *mere* variety. Variety is certainly desirable, but it would be as pedantic to prescribe it at every instant as to forbid consecutive fifths when they sound well.

If by variety we mean specially the monotony of compound tones,

---

[1] Non si debbe anco porre due ò più imperfette consonanze insieme ascendenti ò discendenti l' una dopo l' altra senz' alcun mezo; come sono due Terze maggiori, due minori, due Seste maggiori anco e due minori. Conciosiache non solo si fà contra quello c'ho detto delle Perfette; ma il loro procedere si fà udire alquanto aspro; per non haver nella lor modulatione da parte alcuna l' intervallo del Semitonio maggiore, nel quale consiste tutto 'l buono nella Musica, e senza lui ogni Modulatione ed ogni Harmonia è dura, aspra, e quasi inconsonante [82, 217].

surely the condensation of the partial components into the range of an octave would make a sufficient variation from the compound tone. And the inevitable departure from the justness of the fifths when the music remains consistently in one key would help to assuage the monotony. Consecutive fifths would then partly cease to be fifths, and should in so far be tolerable. Besides, if two consecutive fifths of different pitch involve too little variety and are therefore forbidden, is there not still less variety in the repetition of one and the same fifth? And yet that sequence is unobjectionable.

Thus it appears that the argument is insufficient to explain the effect produced by two consecutive fifths. Some much more decided difference must be the source of the ugliness in question.

We shall see as we proceed that, if the word 'independence,' which has often been used in this connexion instead of the word 'variety,' were properly emphasised and defined, much could be said in favour of the theory. The reader may therefore in the end feel that the variety theory has much in its favour. No doubt those who advocated it felt this underlying justification, but in their expressions they refer only to independence in the sense of variety or *difference* of voices. They did not mean by independence the independence, as distinct from the mere variety, of the voices.

4. *Ambiguity of key or tonality.* One of the theories discussed by Sacchi falls more or less under this head, although it is not quite the same as the form most familiar at the present time. The prohibition of consecutive fifths in this case was held to be due to the great difference between the scales that arise from the bases of the successive chords. Thus the scale of *d* differs from that of *c* in two notes (the two sharps).

But, said Sacchi, why not imagine in the scale of *d* instead of a major, a minor, third (as in the melodic minor scale). Then there would be a difference of only one note, i.e. the least possible difference, between the scales. And, after all, the difference alleged is not really heard, it is only imagined, or at least conceived. The argument surely attributes too great force to the imagination. Besides, the four tones heard (*c–g* and *d–a*) actually belong to one and the same scale and have optimal consonance with one another. How then could they be turned into an offence by mere imagination or rather by the mere possibility of imagining two other notes? "This will certainly in no wise happen : for things imagined, and that can be, never prevail over such as are or are felt" (57, 8 ff.).

Although the theory in question would hardly find a champion nowadays, the argument against the force of imagination is noteworthy. Imagination or the inclination of interpretation is often very important in music, but it is well to be reminded that such a fluctuating and divertible force is not likely to be the cause of a very constant and highly undivertible phenomenon, such as the unpleasantness of consecutive fifths.

The theory of the prohibition advanced by Sacchi himself is the ambiguity of tonality created by the sequence of fifths, e.g. *c–g* and *d–a*.

I have no sufficient reason to refer the second note (*d*) to the first (*c*) or conversely the first to the second (57, 76). I might be told that I ought to refer the second *ut* to the first, because the sound of the first has already taken possession of my ear. But with equal reason I might be told that the first *ut* ought to refer to the second, because all other things being equal, the present sensation, being more lively, is to be given precedence of the past. The suspension is therefore unrelieved; and I am lost in ambiguity, indetermination and suspense between two different notes, each of which, without any difference, can be considered as primary base (*ibid.* f.).

Those who follow this type of theory do not usually admit that all other things are so perfectly equal as Sacchi said. And if they are, is there not an obvious means of overcoming the ambiguity—by indicating the key with all possible precision before the consecutive fifths are introduced? Then the ground of objection to them alleged by Sacchi would be removed. If imagination is an insufficient force to explain the offensiveness in question, we must surely conclude that any such ambiguity is only very slightly more potent. An easy way out of any such difficulty is in constant evidence in every exactly measured sequence of objectively equal intervals of time. The attention may elect to hear this sequence as one or other of various rhythms: ˘´, ˘´, ˘´ ... or ˋ˘,ˋ˘,ˋ˘ ... or ˋ˘˘,ˋ˘˘,ˋ˘˘ ... and so on. But in spite of the great number of possibilities, each of which can be realised at inclination, the ear never remains tortured by ambiguity. It adopts at once a definite rhythm, possibly the easiest under the circumstances. When it has had enough of that one, it 'fluctuates' into another, it may be. If the answer be that such an involuntary solution is impossible in dealing with keys because they are different from mere rhythms, then we must reply that if the apprehension of tonality is at all difficult and not so inevitable as is rhythm, then the mind should receive the sequent chords without any 'thought' of tonality at all,—as an unmusical mind certainly would.

And Sacchi himself excludes the last possible reply at this point by his words :

In fact the two successive fifths not only offend the ear of the erudite in music, but of those even who have no practice in it, so long as they happen to have been born with a good and subtle ear, and pay attention to what they hear. For where one does not attend, which of the dissonances will not pass unobserved and without offending? (57, 16).

The central attitude to chords, whereby the lowest component is the most sonorous, most before the attention, etc., may well be inevitable even to those unpractised in the ways of music; but no one could well suggest that the apprehension of tonality is such an inevitable and 'natural' attitude, requiring no practice and experience. There is some evidence that the Greeks related the pitches of all their notes to one particular note—the *mese*,—at least for the purposes of tuning, if not with some feeling for 'tonality.' In the latter case the functions of their tonic must at least have been very different from ours. Before a certain period of modern music the sense of tonality was much weaker than it is now. In various cases it may have hardly been present at all.

Finally, ambiguity of tonality is by no means uncommon in more modern music at least. Transposition from one key to another is frequently affected by means of one or more chords that are common to both keys. In a familiar piece of music, then, one may not only 'hear' the key that is to be left but one may be able to anticipate that to come, so that the transitional chords may be in a real sense ambiguous. And yet no such horrid effect is thereby produced as is characteristic of consecutive fifths. Considerable dispute is often possible as to which key a short passage of a musical piece really displays; but no specially inartistic effect appertains to such passages. Mere ambiguity of key without any other difference is therefore useless as an explanation of the prohibition of two sequent fifths.

Perhaps no explanation is more frequently offered for the disagreeable effect of consecutive fifths than that suggested by Cherubini: viz. that two parts moving progressively by fifths are moving in two different scales[1]. The reason is obviously insufficient; but it has more force than some critics are willing to admit, when, for instance, three triads in succession are based upon the notes C, D, E, the first having

---

[1] For a very primitive form of this theory—in the writer of the Commentary called *Scholia Enchiriadis*, see 81, 56 f.: "We learn that it is the impropriety of this combination of two different modes or species of the scale, throughout the whole of a composition, which in his view gives rise to the necessity for a free treatment [of the organum]." (Quare in Diatessaron symphonia vox organalis sic absolute convenire cum voce principali non potest, sicut in symphoniis aliis? Quoniam per quartanas regiones non iidem tropi reperiuntur, diversorumque troporum modi per totum ire simul ire nequeunt. 83, vol. 132, 1003; cf. p. 972.)

a major third, and the other two having minor thirds, but each with a perfect fifth, it seems clear that the parts do not all progress in one scale throughout. The upper parts cannot really be in the scale of C, because, as we know, neither a true fifth nor a minor third can, strictly speaking, be based upon the second degree of an accurately tuned major scale. But it may be said that the instrument upon which these chords have been played, is tuned upon the system known as 'equal tempera-ment.' No doubt it is. Nevertheless I contend that, as we tolerate its sharp major thirds and flat fifths, knowing and feeling them to be substitutes for the true thirds and fifths of the genuine scale, so we are accustomed to accept other divisions of the scale, not for what they actually are, but for what they represent.... To me, therefore, it does not seem unreasonable to argue that even with the pianoforte we recognise the equivocal nature of such a progression as that contained in my first example [triads in C, D, E]. But even after this has been granted, the argument that consecutive fifths cause two parts to move in different scales cannot be carried much further. The triads on the third, fourth, fifth, and sixth degrees are all perfectly in tune in the scale of just intonation. Some further reason, therefore, must be sought for the unquestionably ugly result produced when these chords are taken in regular rotation, either ascending or descending (17, 100 f.; cf. G. A. Macfarren (17, 119 f.; 35, 10).

The argument thus properly refuted by F. E. Gladstone contains at the best something very like a dilemma which renders it obscure and confusing. When we hear fifths true to the scale of c major on c, d and e, our musical habit may make us do either of two things : either we are governed by our habit of the major scale on c, and then the intervals are heard as played and the second is not a true, but only an approximate, fifth and is heard as such; or we are drawn away rather by the habit of the perfect fifth, when the intervals will not be heard as they are played, but only as they suggest—in perfect form—and the sequence will lie in no one scale. If the former alternative is valid, the prohibition of consecutive fifths must hold for all fifths, whether perfect or approximate, if the ugly effect persists in spite of the recognised approximation (and that it does persist could hardly be denied). The alternative would only prohibit such fifths as lead to a distortion of the intended scale. A third alternative might claim that we hear all the tones and intervals as played, but that we take the approximate fifths as representing true ones and are disturbed by the distortion of scale thereby implied. This would surely be a needless procedure in view of the fact that even if we take approximate thirds as representing just thirds (minor), we do not thereby feel any distortion of scale.

Of these three alternatives the first seems not only easier and more natural but also most in accordance with the whole system of musical synthesis and apprehension. It is the only one which makes the system of equal temperament musically tolerable for permanent use or at least as—for any individual—the only known system. For the other alternatives presuppose that the pianoforte is only tolerable in virtue of the just scales it suggests, which must have been otherwise ineradicably planted in the minds of all those who use that instrument. That is probably true of those whose experience has made them most familiar with instruments that play just intervals. But for the great majority the intervals their minds apprehend when the piano is played are those actually played, the scale known is the pianoforte scale, and the rules and prohibitions of musical structure are valid for all these approximations.

It is important to notice that these approximations are primarily matters of consonance and its grades. An equal fifth partakes as a consonance of the grade of fusion found optimally in the just fifth. The same holds still more for the dissonances. And as intervals the latter can be learnt in their approximate form; the just form of the dissonant intervals has no such special nature as a fusion as would enable it to draw the ear towards itself. It is a familiar fact that the high grade consonances have this power. It is difficult even to strike the dissonances that lie near these consonances because the voice tends to slip into the easier and more familiar consonance.

A special objection to the key theory lies in the fact that consecutive fifths are only objectionable when they lie between the same voices of the music. That is a primary condition of the phenomenon, but it is by no means a condition of sameness of key. In music of any definite tonality all the voices of any moment are held, and are intended, to be in the same key, whatever that may be. There is no recognised complication of tonality in which a group of keys are considered to be maintained concurrently, one in each separate voice, or pair of voices. Hence the key theory loses its ground entirely.

It is true that in recent compositions parts or groups of parts have sometimes been made to move concurrently in different keys (cf. 23, 138 f.). But that is quite a different matter. Each group of parts is still subject to the fundamental rules of harmony, although as between the separate groups the claims of these are largely ignored.

5. *Want of relationship.* The theory advanced by Gladstone is "that consecutive fifths are generally more or less offensive in proportion to the want of relationship, or otherwise, existing between the chords which produce them." He cited Kollmann as probably the first writer

who propounded this idea[1]. Unfortunately his case was spoilt at the very outset by his admission that there is as little relationship between the inversions of chords on successive notes of the scale as between their original positions; and of the inversions a succession of six-three chords is quite admissible. The objection attaching to the six-fours does not alter this fact in the least, nor its destruction of the theory of relationship. We may, therefore, expect the frequency with which consecutive fifths are admitted between tonic and dominant chords and *vice versa*, or between tonic and subdominant chords and *vice versa*, to have some other cause. We cannot interpret this frequency as due solely to the high degree of relationship between the chords in the sense of relationship implied when we say that the triads on C and D are totally unrelated. Gladstone admitted besides, that he had "not met with any specimens of consecutive fifths in which the roots of the chords *rise* a third (except where a sudden change of key occurs)" (17, 104), and evidently had found only two cases of a fall of a third—tonic to submediant.

Gladstone proceeded to note that the objection might be raised that this argument ought also to apply to fourths, thirds, and sixths; and recalled in reply that the movement of fourths *is* placed under various restrictions by the laws of counterpoint, and that even two major thirds in succession are still forbidden in the strictest style of two-part writing (17, 105). But however interesting this extension of the basis of argument may be, it is perfectly obvious that the unexplained relaxation of the prohibition in these cases only makes the theory of relationship the more impossible.

6. *The nature of the interval itself.* The study of the connexion between consecutive fifths (and octaves) and the harmonic relationship of chords has been renewed by Shinn (58, 265 ff.; 59). But Shinn does not attempt to explain the prohibition of successive fifths and octaves by the lack of harmonic relationship between the chords in which they stand, but by the intrinsic character of the intervals of the

---

[1] It was also the basis of Pearsall's (1795–1856) explanation who wrote that "consecutive great thirds and perfect fifths are evidences that some harmony has been sprung over which ought to have been introduced by its characteristic note, as forming the natural link of relationship between these intervals." When they are not evidences of such a spring, "they carry with them an awkwardness of progression which ought to be avoided." They display "a want of freedom and a clumsiness, unacceptable to any musical ear" (49, 25). But not even strong disapproval forms a logical complement to an incomplete theory.

octave and the fifth themselves. For the octave Shinn adopts the explanation that may be termed traditional. The bad effect of a 'hidden' octave and *à fortiori* of consecutive octaves is "the weakness which is produced by the correspondence in sound of the two outside parts (approached in this manner)" (58, 268). This statement revives the explanation by want of variety discussed above. The bad effect of a hidden fifth, and *à fortiori* of consecutive fifths is due "to the bareness of the interval" (58, 268, 280).

This expression may be taken either as a variant upon the "weakness" of the octave or as a new kind of reason that has not been advanced by any other theorist, as far as I am aware. It makes perhaps some approach towards the notion of consonance as approximate unity. But obviously it is now an aspect of the interval (or fusion) of the fifth that displeases, whereas in the octave it is the correspondence in sound of the two tones that make up the interval—the so-called identity or similarity of octave-tones. Such a difference of causes could hardly be acceptable.

It is not easy to give a just systematic place to Shinn's exposition. In some ways he suggests the first theory of this chapter. He says, for example (58, 263) :

> Combinations and progressions which were formerly regarded as painfully crude, harsh and ugly, have, by familiarity, lost these characteristics, and become both piquant and pleasant; while others, which had hitherto produced pleasure, now seem commonplace in comparison with the poignancy of less familiar but more forcible ones. In connexion with this matter, the important point to be recognised is, that no change has taken place in the progressions themselves, but it is the ear of the listener which has changed, owing to the influence of a change in his musical environment (58, 263).

Elsewhere he speaks of the "so-called objectionable effect of consecutive fifths" (p. 284).

On the other hand he points out that this bad effect (whether so-called or not) is "almost invariably neutralised by the harmonic relationship which exists between the chords forming such fifths" (*ibid.*). The bareness of the fifth, we are to understand, is somehow annulled or enriched. But Shinn neither explains why the fifth is a bare interval—I suppose it just sounds so—nor does he show how chord relationship removes this bare character from the interval. In fact he is not always quite faithful to the explanation by harmonic relationship and abandons it in part in favour of "the effect of musical strength which is characteristic of (such) progressions" of the two voices con-

cerned by a fourth and a fifth respectively (58, 277). It is not entirely
a matter of "harmonic relationship (or root progression), but partly
also of the movement of the outside parts by almost equal intervals."
The latter acts even without any special harmonic relation.

Nevertheless Shinn's discussion is, as we shall see later, probably
the most 'philosophical' one that has so far been given.

7. *Want of balance.* A suggestion made by G. A. Macfarren is
worthy of mention, although it has not been worked out into a definite
theory as far as I am aware. "When a passage of harmony in any
number of parts has two notes made so very much more prominent
than the rest, as is the case in the duplication of those two at the expense
of the others, the other portion of the harmony is enfeebled, and the
balance is destroyed" (17, 118). This is exemplified in the case of
successive octaves, when two notes mutually reinforce each other and
so become particularly prominent over against the rest of the score.
The same does not hold for the doubling of a voice, even although the
relation is now merely two to three instead of two to two; for here the
doubled part is meant to be specially prominent.

In this restricted form the theory of balance is not very significant.
For the balance in question is chiefly a balance of mere strength. An
overbalance of strength can easily be produced in music and often
occurs not only by mere accident and through the imperfect technique
of performers, but through want of finish on the part of the composer.
On none of these occasions could it well be said to be so strikingly
unpleasant as to justify the view that consecutive octaves (and fifths)
are forbidden so strictly because of the disproportion of strength they
produce.

CHAPTER XII

THE SYSTEM OF FACTS REGARDING CONSECUTIVES

THE preceding chapter contained a review of the explanations that have been attempted for the prohibition of consecutive fifths. We have noticed how the various theories try to set the prohibition into relation to facts of a similar kind so as to obtain some indication of systematic coherence in the explanation. None of the systems of facts thus suggested is very satisfactory or convincing; and none of the theories can possibly be held to be successful. It is extremely doubtful whether anyone who has thus far reflected on the problem of consecutive fifths has felt that more than interesting suggestions towards an explanation have been reached.

"Often and often have I thought," said G. A. Macfarren, "it would require the entire knowledge of a physicist to be able to probe this subject to its foundation" (17, 119). But the day when physical science may be expected to solve such a problem is now definitely past. Even Helmholtz, whose basis of explanation might well seem to many to be physical, was perfectly well aware that the ground of explanation of all musical phenomena must lie within the phenomenal stuff of sound itself; it dare not be merely physical (20, 231f., 368). No doubt the prominence, or at least the great propinquity, of the physical throughout his exposition prevented many of his followers from giving sufficient heed to the psychical or, if you like, to the phenomenal aspect of the problems of music. Besides the difficulty and apparent obscurity of the psychical itself made them only too eager to seize upon any plausible excuse for evading the study of its elementary aspects. Such an excuse was not only given, but even emphasised by Helmholtz.

The system of scales, modes, and harmonic tissues does not rest solely upon inalterable natural laws, but is also, at least partly, the result of esthetical principles, which have already changed, and will still further change, with the progressive development of humanity (20, 235).

This proposition, he said, was "not even now sufficiently present to the minds of our theorists and historians." But ever since then, at least, it has been decidedly obstructive in its effect upon their minds. It was a most unfortunate dictum. For the opposition implied in it between natural laws and aesthetical principles strongly suggested that the latter are merely arbitrary conventions, as is more or less, for example, the fashion of clothes in any year. After quoting this dictum Prout wrote :

> While, therefore, the author [himself] follows Day and Ouseley in taking the harmonic series as the basis of his calculations, he claims the right to make his own selection, on aesthetic grounds, from these harmonics, and to use only such of them as appear needful to explain the practice of the great masters (52, 1st ed., 1889 iv).

And many others besides Prout could be quoted to the same effect.

But an aesthetical principle is not the sort of thing that men for centuries in vain seek to explain. So hidden a cause is rather an aesthetical law,—which is just as much law as is any physical uniformity. And it can no more be laid aside in this arbitrary way than an ethical standard can be suppressed whenever you think it will not approve of what you choose to do.

A noticeable feature of these attempted explanations of consecutive fifths is their fragmentariness and isolation. Most theorists give only a short statement of what seems to them to be an easy and obvious reason for the prohibition of octaves, namely, the disturbance they produce in the balance or in the melodic distinctiveness of the parts. And some theorists refer to the minor restrictions placed upon sequences of fourths (and even of thirds) in confirmation of the different theory they offer for fifths. It may seem to many minds quite satisfactory to have one solution for the octaves and a second for the fifths and other intervals. Difficulties that are allowed to slumber, of course make no attack. But there still remain the few intervals that are not prohibited in succession at all. No one who has a keen sense for the systematic logic of a theory can long remain satisfied with such work. And so the problem of consecutive fifths remains to-day without any recognised solution.

Once the psychical ground of the phenomena of music has been recognised, it may seem to be an inevitable consequence that no satisfactory or convincing explanation of such phenomena can be given. There are many who think the appeal to the subjective judgment necessarily unconvincing. In his introduction to his account of Rameau's doctrines D'Alembert wrote :

> Here must not be sought that striking evidence that is peculiar to works of geometry and that is so seldom met with in those in which physics mingles. There will always enter into the theory of musical phenomena a sort of metaphysics that these phenomena implicitly suggest and that brings thither its own natural obscurity; in this matter we must not look for what is called *demonstration*; it is much to have

reduced the principal facts to a system well linked and well pursued, to have deduced them from a single experiment, and to have established on this so simple basis the best known rules of musical art. But, on the other hand, if it is unjust to exact here that intimate and unassailable persuasion that is produced only by the most vivid light, at the same time we doubt if it is possible to throw a greater light upon these matters (9, xiii f.).

Since D'Alembert's time even the demonstrations of physics, that have become so numerous as to be a sort of standard for all sciences, have been subjected to such searching examination as to make some minds incline to see in them only a complete description of events. No doubt there is much more involved in them than this. But that more is itself the source, not of a superiority of physics to the science of the foundations of music, but on the contrary of a kind of philosophical inferiority. For physics implies the positing of many types of real entities—the substantial basis of the phenomena that are so perfectly described. At least all but a very few thinkers allow this feature to enter freely into their physical constructions. Only a few extremists— shall we say?—such as Ernst Mach and Bertrand Russell, have attempted to claim that physics as a science may be construed without any such postulations, but merely by description or by classification (of phenomena) as a fundamental process.

The science of music, however, has as it were its whole perspective in converse form. Its facts are obviously phenomenal; it not only begins with the completest description, but, in the opinion of many, it necessarily ends there too. For music, they hold, is entirely phenomenal. Of course, when the basis of explanation is carried back into the physical realm, as—after sufficient description and explanation of the phenomena themselves—it properly may, our knowledge of the basis of music then goes beyond the bounds of the phenomenal realm. That, however, is not the point at issue. Those who say the basis of music is entirely phenomenal mean that its whole task is necessarily mere description. Description and classification, and the study of sequence and of dependence amongst phenomena cannot, they think, lead to any knowledge of phenomena that could show them to be not wholly phenomenal.

Only very few venture to claim that the science of musical phenomena may gain knowledge of these phenomena that shows them to be more than phenomena, to be at least partly real, entities independent of our minds that do not necessarily reveal themselves completely and finally to us at the first glance or after any amount of inspection. In so far then, as the science of music passes so rarely or never over the frontiers of the phenomenal, its descriptions can proceed with fewer questionable or uncertain assumptions, and may be looked upon as more completely defensible in a logical sense than can even the work of so highly successful a science as physics.

However that may be, there is no doubt at all nowadays that the science of phenomena can attain to as complete description of its objects as any science, and can thereby compel conviction as completely. No doubt the differences to be described are often very subtle; but they may often be clear and easy to distinguish. In any case we must not nourish false expectations. Phenomena cannot, for example, be magnified with microscopes. We are at the very outset already at the limits of possible magnification. But apart from this sort of thing, the methods of a science of phenomena are as reliable and—to those whose minds are open to conviction—as convincing as are those of any science of nature.

There are indeed very many at the present time whose minds for various reasons have closed completely against this idea that a sufficient science could be made of mere phenomena such as is the heard stuff of music. They would not deny that an art could be raised upon this basis. But art they may feel as subjective and variable with the caprice and inclination or even with the 'personality' of each man. Nevertheless we must insist upon it that the more an art is studied, the more it is felt to be a realm of order and coherence. No doubt in a science of art we are dealing with the finer, more intimate, issues of events, and not so much with the great lines of Nature's efforts. But in arts there are also broad beams of construction. And the natural sciences have all already come into contact with the subtlest and finest issues of their objects, so that this difference between the sciences of Nature and of Art has no longer any effective validity with reference to their dignity as systems of knowledge.

The first task that presents itself in every problem of the foundations of music is to describe the phenomena as completely as possible. Preliminary to this is the effort to get all the phenomena together. That must be done by starting from the phenomenon that first raised the problem,—for example, our problem of the prohibition of consecutive fifths,—by searching for all phenomena in any degree similar to this striking one, and by endeavouring to find a systematic arrangement and description that will incorporate them all.

That is the logical status of the method of solution. It may in some cases be the method of discovery as well. As a matter of fact the systematic arrangement now to be expounded was first suggested by an attempt to reduce the rules for part-writing given in E. Prout's *Harmony* (52) to comprehensive and facile form by making a table of the objects shown by the rules to be of chief importance—octaves, fifths, fourths, sevenths, seconds, and ninths—along with the recurrent factors that seemed to modify their admission or prohibition,—e.g. the different voices, the kind of motion, etc. This effort suggested its own extension and completion to what seems highly probable as at least a close approximation to the system of facts of which consecutive fifths form a part and from which a satisfactory explanation of their prohibition may flow. This system of facts seems, moreover, to fall into place as an extension of the system of facts regarding the foundations of music already expounded. Not only so but it seems also to renew their ground in an independent manner, which may lend great weight both to the facts as already described and to the systematic description and explanation they have received.

Thus the series of intervals just mentioned may be completed by the addition of thirds and sixths. Then we have octaves, sevenths, sixths, fifth, tritone as diminished fifth or as augmented fourth, fourth, thirds, and seconds. We shall omit consideration of the so-called prime or of unison for the present (cf. below, p. 112). These are all the intervals smaller than the octave. If we arrange them in their order of fusion, we get: octave, fifth, fourth, thirds and sixths, tritone, sevenths and seconds. For the study of consecutives in general this order is of much greater importance than is the former.

Of the circumstances that modify the prohibition of consecutives the series of voices is one of the most important. Let us begin with the consideration of four-part harmony. The bass, as we have already seen, is the naturally predominant voice. The next in order is the soprano, partly because it is the other outside voice, and partly because it usually bears the most important, if not the only (coherent or thematised) melody in harmonic music. It is the only voice that is often claimed to be more noticeable in a single stationary chord than is the bass (cf. p. 51 ff., above). There is no obvious distinction in importance between the two other voices. This grading of the voices is confirmed and greatly strengthened by the way in which the stringency of the rules of part-writing is relaxed on occasion in relation to the different voices.

But each interval must necessarily involve two voices at once, so that the series—bass, soprano, tenor or alto,—has to be squared with itself, so to speak. The grading that ensues is : (1) bass-soprano, (2) bass-alto or bass-tenor, (3) soprano-alto or soprano-tenor, (4) alto-tenor; or, as we shall often find it convenient to write them : B–S, B–A and B–T, S–A and S–T, A–T.

When this series is correlated with the series of intervals just stated, Table I results. No definite preconceived idea determined the form of it. The idea was merely to arrange the chief objects referred to in the rules of part-writing as given by E. Prout (52, 25 ff.) and their chief relations so that any system implicit in them might become patent.

TABLE I

Consecutives (preliminary system)

Showing relations between (1) the stringency of prohibition (Forb. – = almost strictly forbidden; forb. = forbidden with exceptions; forb. – = with more exceptions; + = allowed) and (2) the grade of fusion or consonance, and (3) the prominence of the voice-parts. Based upon the formulations of E. Prout (52, 25 ff.).

	Octaves	Tritone to fifth	Fifths	Fourths. Tritone to fourth	Thirds or sixths	Sevenths	Seconds or ninths
B–S	Forb. –	Forb.	forb.	forb.	+	forb.	Forb.
B–A	Forb. –	Forb.	forb. –	forb. –	+	forb.	Forb.
B–T	Forb. –	Forb.	forb. –	forb. –	+	forb.	Forb.
S–A	Forb. –	forb.	forb. –	+	+	forb.	Forb.
S–T	Forb. –	forb.	forb. –	+	+	forb.	Forb.
A–T	Forb. –	forb.	forb. –	+	+	forb.	Forb.

Tritone is the diminished fifth or augmented fourth, as the case may be; it is not reckoned as a fourth or a fifth when it follows a fourth or a fifth respectively. The Table has been arranged so as to bring out a grading in the stringency of prohibition from below upwards and from side to side.

The rules upon which this Table is based are the following (52, 25 ff.) :

(1) No two parts in harmony may move [in unison, or] in octaves with one another. There is one exception to the prohibition of consecutive octaves. They are allowed by contrary motion between the primary chords [tonic, dominant, sub-dominant] of the key, provided that one part leaps a fourth and the other a fifth.

(2) Consecutive perfect fifths by similar motion are not allowed between any two parts. They are, however, much less objectionable when taken by contrary motion, especially if one of the parts be a middle part and the progression be between primary chords [T.D.Sd.]. This rule is much more frequently broken by great

composers than the rule prohibiting consecutive octaves. Consecutive fifths between the tonic and dominant chords are not infrequently met with.

If one of the two fifths is diminished, the rule does not apply, provided the perfect fifth comes first.... But a diminished fifth followed by a perfect fifth is forbidden between the bass and any upper part but allowed between two upper or middle parts, provided the lower or occasionally the upper part moves a semitone.

(3) Consecutive fourths between the bass and an upper part are forbidden, except when the second of the two is a part of a fundamental discord [whose intervals are a major third, a perfect fifth, and a minor seventh from the generator[1] (39, 94)] or a passing note,—i.e. a note not belonging to the harmony. Between any of the upper parts consecutive fourths are not prohibited. They are sometimes found between the bass and a middle part; but even these are not advisable.

(4) Consecutive seconds, sevenths and ninths are forbidden between any two parts, unless one of the notes be a passing note.... There is one important exception to this rule to be found in the works of the old masters. Corelli, Handel, and others sometimes followed a dominant seventh by another seventh on the bass note next below.

There is some difference of opinion amongst authorities as to the special digressions from the prohibitions that are admissible. But if we take account chiefly of the existence and degree of freedom of exceptions we may look upon Prout's rules as relatively valid.

Thus Parry (45) says that "there are so many consecutive sevenths to be found in the works of the greatest masters, and that, when they are harsh, they are so obviously so, that the rule prohibiting them seems both doubtful and unnecessary." Here the point of view is mainly practical. The ugliness of the sequence is really admitted, although in a restricted form; and so it appears in our Table. Text-books of harmony evidently find it unnecessary to state that consecutive thirds and sixths are unobjectionable. But the fact is of the greatest importance as a datum for theoretical work for all that. It is just these obvious facts that no one mentions that are often the keystone of a successful theory of such phenomena.

There is evidently a system inherent in the Table. For we see that the two outside columns contain nothing but Forb. or Forb. − , i.e. strict or almost strict prohibitions for all voices. In the second column the prohibition is relaxed a little in the lower three pairs of voices (forb.). In the third column that relaxation, and perhaps even a little more of it (forb. −), holds for all the pairs of voices except B–S. In the fourth column further progress is made in the same direction; the sequence is admitted for the lower three pairs of voices. In the case

---

[1] The 'generator' is the lowest note of the so-called root-position of the chord.

of the thirds and sixths there is perfect freedom. But the prohibition comes into force again in a restricted degree for the sevenths, and is finally complete for the seconds and ninths. The grading of the voice-pairs that makes this system possible is identical with the grading deduced above from the experimental and general facts of analysis of chords.

The most significant and suggestive feature of the Table, however, is the sequence of the intervals seen in the top horizontal column. This sequence is very nearly the same as that already given for the grades of consonance of intervals that lie within the octave. And the result suggests strongly,—so far at least as this Table shows the situation,—that the degree of stringency of the prohibition of consecutive intervals of the same species depends (1) upon the grade of consonance or dissonance, and (2) upon the prominence of the two voices that constitute the intervals in question. The Table thus takes proper notice of the fact that the consecutive intervals must both lie between the same two voices.

A special problem is created in the Table by the tritone. As a so-called diminished fifth it is not forbidden when it follows, but only when it precedes, a perfect fifth. The same holds in so far as it is reckoned as an augmented fourth in connexion with a perfect fourth. The prohibition of the tritone when the fifth follows it, seems to be stricter even than the prohibition of two fifths. This extreme difference between the two successions shows that we are here not dealing with consecutive intervals of the same species. When the augmented fourth or diminished fifth follows, it is not a fifth or a fourth at all, but another kind of interval—at least as far as the system of facts represented in the Table is concerned. It is for that reason it has been classified in the Table as a tritone. Thus the problem of the tritone reduces itself to the single case in which it precedes either a fifth or a fourth. This is obviously not an instance of consecutive intervals of the same kind. In respect of the fifth alone it belongs to the case of 'hidden fifths.' Otherwise the succession of tritone and fifth belongs to the class of problems that includes all such questions as : under what circumstances may any two intervals of *different* species follow one another in the same voices?

We may, therefore, remove the tritone from its present position in the preliminary Table of consecutives and replace it properly, having regard solely to the degree in which consecutive tritones are avoided

or forbidden in part-writing. Text-books of harmony contain no statement regarding consecutive tritones, either as diminished fifths or as
augmented fourths. But in dealing with modulation a practical opportunity occurs of presenting information on this subject. Thus P.
Tchaikovsky says (76, 70 f.):

> There are also sequences in which every chord constitutes a modulation. They
> are those in which dominant seventh chords or other chords resolving into the
> tonic succeed one another, always falling a fifth or rising a fourth, as in a sequence
> within the limits of one key. In such a sequence each chord resolves into a chord
> which itself demands resolution and forms at the same time the resolution of its
> precursor.

He then gives progressions containing five to eight tritones in
succession, diminished fifths alternating with augmented fourths. Of
course there is some difference between these two intervals in their
musical significance. But in respect of their fusion and even in respect
of their specific nature as intervals (proportions of volumes) there is
practically none. In one of Tchaikovsky's examples there is even a
series of eight simultaneous *pairs* of tritones (chords of the diminished
seventh), one tritone lying between the two outer voices, the other
between the two inner voices. Prout (52, 1st ed., 162) gives an example
from Bach's Chromatic Fantasia in which successive tritones abound.

We may, therefore, look upon successive tritones as being more
freely admissible than consecutive sevenths, major or minor. The
diminished seventh, which may be run in succession (52, 242) is practically,
and from the point of view of fusion, quite the same thing as the major
sixth, which is not restricted at all. No doubt the musical affinities of
intervals that are apprehended as diminished sevenths will call for a
different treatment of them from that of intervals apprehended as major
sixths. But it is clear that in dealing with consecutives we are not
concerned with such special apprehension of musical setting and relationship, but with a more fundamental matter that appertains to the
intervals in question almost in any setting, so long as they lie between
the same voices.

We may, therefore, place the tritone in an amended Table of
Consecutives between the thirds, sixths and the sevenths.

This brings the final Table into much greater conformity with the
experimentally established grading of fusion of the different intervals.
In fact so far as the differentiation of our Table shows, in which the
different thirds or sixths or sevenths, etc., are not distinguished, the

conformity is complete. The most frequent grading of fusion that experimental research has as yet shown is (77, 104): O, 5, 4, III, 3, VI, 6, T, II, 7, 2, VII (cf. above, p. 16).

TABLE II

Consecutives (*Final System*)

Showing relations between (1) the stringency of prohibition (Forb., Forb. -, forb., forb. -, + or allowed), and (2) the grade of fusion of any interval, and (3) the prominence of the voice-parts in which the interval appears.

	O's	5's	4's	3's & 6's	T's	7's	2's & 9's
B–S	Forb. –	forb.	forb.	+	forb. –	forb.	Forb.
B–A	Forb. –	forb. –	forb. –	+	forb. –	forb.	Forb.
B–T	Forb. –	forb. –	forb. –	+	forb. –	forb.	Forb.
S–A	Forb. –	forb. –	+	+	forb. –	forb.	Forb.
S–T	Forb. –	forb. –	+	+	forb. –	forb.	Forb.
A–T	Forb. –	forb. –	+	+	forb. –	forb.	Forb.

The Table shows that the grades of fusion from greatest consonance to greatest dissonance—in relation with the relative prominence of the pair of voices on which the interval in question rests—give rise to a system of preferences or prohibitions of a very well graded kind (cf. p. 103, above).

And the following conclusion may be drawn. If due consideration is given to the prominence of the pair of voices that bear the interval in question, it appears that the immediate repetition of an interval in the same voices is the more offensive the greater the consonance or dissonance of that interval. The point of minimal unpleasantness or of maximal pleasantness (as the case may be) in the series from greatest consonance to greatest dissonance lies amongst the thirds and sixths. These intervals may, therefore, be held to be fusionally neutral.

This inference differs a little from the prevalent attitude towards the thirds and sixths. They are nowadays ranked among the consonances. They are, of course, certainly not dissonances. But, on the other hand, they are perhaps not really consonances either.

It is a familiar fact that the ancient Greeks did not include them amongst their consonances, which were octave, fifth, and fourth, alone, and stated in this order by Aristoxenus and by Ptolemy (cf. 66, 38, 58). This fact has often been interpreted as indicating that the Greeks considered the thirds and sixths to be dissonances, as we now understand this term. But that may not be taken for granted. The system indicated in Table II suggests strongly that the thirds and sixths may not have been included amongst the consonances by the Greeks because they are not appreciably *positive* degrees of consonance[1]. That we find them highly pleasant and characteristic is not at all inconsistent with the correctness of this estimate.

Of course the mere fact that we rank the thirds and sixths after the fourth in the grading of fusions that lead from greatest consonance to greatest dissonance, implies nothing at all as to whether these intervals are consonances or dissonances. And the experimental evidence regarding the very slight percentage difference between the grades (of approximation to the impression of a single tone) lower than the fourth, and the similar evidence regarding the variations in the serial arrangement of the grades of fusion show that at least there is no clear division between these lower grades. So if a minor seventh is a dissonance, a minor third can hardly be a strong consonance; nor can even a major third. There must be a point at which dissonance passes into consonance. Logically that point may be a vanishing point, of course. But even then the lower consonances,—if we suppose the thirds and sixths to be positive consonances,—must have a very low degree of consonance to be so slightly different from the lesser dissonances and so often confusable with them in respect of fusion. It, therefore, seems probable that the grades of fusion including the thirds and sixths may properly be considered to be neutral.

Thirds and sixths, then, are neither distinct dissonances, nor are they distinct consonances. And the Table of Consecutives gives us a very strong reason for accepting this description. For the treatment there shown to be accorded to thirds and sixths is distinctly different from that accorded to the extreme consonances and to the extreme dissonances.

---

[1] Cf. Gevaert (14, 102): "Let us notice first that the meaning of the terms has been modified in the course of time. We translate *symphonia* by consonance, *diaphonia* by dissonance. So did even the Romans in the Augustan age, always attaching to these words another idea than we do. The fundamental difference distinguished by the ancients between the two kinds of intervals is that in symphony the sounds fuse to a perfect unity, whilst in diaphony they maintain their individuality and detach themselves in some way from one another. In this respect our impression does not differ appreciably from that of the Greeks. For us too the thirds and sixths have a clear cut character that is lacking in the fifth and fourth. We notice the same clearness in the second and in the seventh; and from this new point of view we find it possible to let the ranking of these two intervals in the same category as the thirds and sixths pass." Gaudentius was the first to admit the major third (and also the tritone) amongst the consonances. To the former inclusion we now generally agree, but only with special effort to the latter (cf. 14, 99; 66, 71 f.).

CHAPTER XIII

THE REASON FOR THE PROHIBITION OF CONSECUTIVES

THE question which next arises is the one from which all previous writers on the subject have started : why are consecutives offensive? Why are all these consecutives offensive, each in its degree?

The system of facts we have discovered in the preceding chapter does not answer the question directly. It only arranges the objective facts with which any answer to the question must reckon. It indicates that the solution must not only be the same for octaves as for fifths, but it must even be the same for both dissonances and consonances. The only thing common to them all is some degree of fusion. The repetition of a high grade consonance or dissonance introduces a new and special feature that is unpleasant. Thus our task must now be to show a basis for this feature and to form a theory as to its nature which will adequately justify on conceptual grounds the unpleasantness of the effect we hear.

The problem may be approached in two ways. In the first, experimentally, we might present a series of observers with a systematically varied complex of consecutive pairs, and ask for direct observation and description of the feature of each that is unpleasant or more or less preferable. This course would certainly not be successful in such a simple form, although it would be of great value for the grading of the intervals on the basis of their preferability. Consecutive intervals of the same species have been considered and compared and reflected upon for centuries already without even any indication of agreement having been reached as to what it is in the sequence that is directly unpleasant. Even those who have written treatises on the subject have hardly done more than guesswork upon the problem, except in so far as they attempted to infer a basis of unpleasantness from the system of facts gathered round the central object of inquiry.

The situation in this particular aspect of it is similar to that of the theory of consonance and dissonance. In spite of the fact that the ancient Greeks and the older writers of the modern era had defined consonance as the mixture or blending of two tones into one, that direct description did not receive in the more modern explanations by the relations of the harmonics the central importance due to it. It was only restored to its proper position by the critical studies of Stumpf. And even Stumpf could not go beyond this amount of direct description, already attained by the Greeks, to say more definitely and decisively how the fusing tones interpenetrated one another so as to approximate to the effect of a single tone. If it was almost impossible in this case to proceed beyond the terms of direct description to an adequate theory of the basis of the phenomenon in the sensory stuff of the tones themselves, how can we expect by direct observation to win a theory of the bad effect of consecutives, seeing that even the direct description of that effect has not yet been obtained?

Evidently the work of observation and description must be facilitated by the discovery of definite alternative questions, to be answered by the comparison of minor differences. In other words, we must learn how to instruct the observer so as to make description easier for him by directing his attention precisely towards possible special features of consecutives. We must expect a properly instructed and careful course of systematic observation to confirm any inferences as to the basis of the prohibitions that may otherwise be gathered.

For that is the other way of approaching the problem. By enlarging the system of facts in which consecutive fifths stand, we have already obtained a much better formula for their prohibition than we could have obtained from a study of them alone. The fifths are forbidden because of something that emerges from pairs of highly positive or negative fusions. Perhaps if we enlarge in turn the system of facts of which consecutives form a part, we may attain some still more specific formula. Armed with this, we could return to the work of direct description with some hope of obtaining a definite answer to a definite question. A probable further system of facts suggests itself in the well-known counterpart to consecutives—the prohibition of single intervals, commonly termed hidden octaves and fifths.

But in the facts already before us there is an aspect that calls for some notice, although it may at first glance seem trivial and obvious : that the intervals to be prohibited must lie between the same voices. Of course that is not equivalent to saying that they must lie somewhere; they might lie between different voices, and when they do so, they give rise to no feature that is objectionable. Evidently when we listen to music in several parts, our attention—even in music that is predominantly harmonic—runs *along* the voices, as it were, noticing the series of relations that emerge between the successive tones of each pair of voices. It does not connect into systems one relation between one pair of voices, a second relation between another pair of voices, a third relation between a third pair, and so on[1]. The systems are rather those that actually present themselves serially. The relations in question, however, are fusional, or, as Hullah said, perpendicular relations. Only, chords are not apprehended—even in music that is predominantly harmonic—as unanalysed wholes; the apprehension is not fusional or perpendicular throughout the chord as an undivided unity. Analysis breaks this whole into parts of two voices at least; these are the units of fusional or perpendicular apprehension.

In polyphonic music this much is also undoubtedly true. Only here there enters another factor that justifies the contrary term 'horizontal apprehension,' namely the distinctively melodic or thematic treatment of each voice. The figures, forms, and phrases of melody are maintained in each voice over and above the restrictions that are placed upon their progressions by the fusional aspects of pairs of voices. It is not, then, so much the case that harmonic music has introduced a feature not yet present in polyphonic music; but rather in the latter there is present in highly cultivated form a feature which prevents the harmonic relations implicit within it from coming into prominence. No doubt, too, this suppression prevented these relations from being specially cultivated. But, whether cultivated or not, they are essentially present in both types of music.

Thus we obtain some closer specification of the relations between synthesis and analysis in music generally. In listening to music in several parts we do not apprehend the fusions of chords in so far as they approximate to the balance and symmetry of a single tone as a whole mass. Our attention is always, up to a certain degree, analytic. We notice always the relations between pairs of voices. And to do so we must be able to maintain the proportions of the volumes as defined by each pair of voices in the forefront of our attention. For that purpose analysis is necessary.

Now we have a perspective from which to judge the generalisation attained from Table II. A succession of high grade consonances or dissonances is very unpleasant; it is offensive according to the degree of consonance or dissonance (or of fusion positive to the neutral grades

---

[1] So the crossing of parts will obviate consecutive fifths that appear when the parts are in pianoforte score (cf. 52, 104 t. for example). But such voice-leading will, of course, require the support of a difference in blend between the voices, as in choral or chamber music.

or negative to them) and to the prominence of the pair of voices concerned. In short, prominence of high or low grade fusion disturbs; neutral grades of fusion do not disturb. Disturb what? Only one answer suggests itself : they disturb (the analysis or the set of attention required to maintain) the usual flow of presentation of relations between the pairs of voices. The horizontal view, so far as it is generally attained in music, is disturbed by the undue prominence of the perpendicular relation between the voices. Either the voices interpenetrate too much in successive pairs so as to cut off the connexion between the two tones of either voice (thus the connexions *c–d* and *g–a* are broken in consecutive fifths on *c* and *d*); or the voices disrupt from one another too markedly and thus also break the connexion unduly within each voice. Neutral grades of fusion alone do not in succession break this even flow of analytic concentration necessary for the appreciation of the greater works of music[1].

We cannot go into the further aspects of this formulation at once. We must await the systematic arrangement of the facts included in these further aspects, keeping the formulation before us as a hypothesis to be tested and enriched. But it is at least evident now why consecutives are forbidden only in connexion with a movement of the voices. The repetition of any interval without any change of its pitch would in no way affect the apprehension of the sequent tones of each voice. For as nothing has changed, the attention has for the moment an easier task than usual. On the other hand a succession of unisons tends to betray the analytic attention into losing hold of the individuality or duple nature of the voices that thus temporarily coincide. One unison, however, is not disturbing so long as the voices are felt melodically to converge and to coalesce; and if the next chord is suitable, they will be felt to separate and to diverge again (cf. below, p. 130).

---

[1] Descartes explained the prohibition of consecutive octaves and fifths thus: "Ratio enim quare id magis expresse prohibeatur in his consonantiis quàm in aliis, est quia hae sunt perfectissimae; ideoque, dum una ex illis audita est, tunc plane auditui satisfactum est. Et nisi illico aliâ consonantiâ ejus attentio renovetur, in eo tantùm occupatur, ut advertat parum varietatem et quodammodo frigidam cantilenae symphoniam. Quod idem in tertiis aliisque non accidit: immò, dum illae iterantur, sustentatur attentio, augeturque desiderium, quo perfectiorem consonantiam expectamus" (11, 132). The comparison with the case of thirds is interesting. But it serves only to bring out the older point of view which concentrated on the perfect consonances, and not the modern point of view in which the thirds play the more essential part. Descartes' theory is of the 'variety' type. That, however, is true only as an approximation towards what variety makes possible and what is attainable in some cases (e.g. with thirds) even without variety, namely continuity of melody.

It must also be now clear that concurrence of voices in fifths or fourths is only tolerable in a primitive stage of music. There the homophonic interest is almost the only one present in the music. Polyphonic relations have either not yet been attained at all or only on rare occasions, so that even when they do occur, the ready dispositions of the hearer's mind will not easily yield to any unpleasantness they may bring when consecutives appear. Both singers and hearers intend and know the concurrent voices to be the same melody. No doubt they have to will, or to attend to, the melodic continuity more energetically when they use consecutive fifths and fourths than when they use octaves or a bare melody. But they may be quite willing to do so for the sake of the variety thus attained, until further variation and closer attention show them that the bad effects thus ignored have no compensating power to please by heightening contrast; or that, if they have this power, the systems of variation made possible by it are so small and weak as compared with the systems of variation admitted by changes of consonances, and especially by the use of the lower degrees of consonance, that they are not worth while, or are not profitable lines of development, and so are best barred out altogether. Hence their gradual disappearance from music as it progressed towards the form and style of distinct polyphony, and their vigorous prohibition until it had developed enough to allow of their re-introduction amongst many minor systems of variation in a way that does tend to enrich the structural potentialities of music.

On the other hand the octave does not make melodic continuity at all difficult to maintain so long as the presence of the intention to such continuity has been made evident or so long as the intention towards melodic diversity has not been declared. The reason for this freedom is not so much the fact that the octave is the first harmonic of a fundamental, whereas the fifth is the second. For that should only establish a gradation of difficulty, as it does perhaps in the primitive mind, not a difference between pleasure and offensiveness, as it does in our music. The reason is rather that in the systems of intervals of our music or in our tonality the octave is the absolute basis of reference of all intervals, and is so because of the fact that the increase of an octave means the decrease of volume by half, and because this difference does not alter or distort any pattern of volumic proportions (cf. above p. 72 ff.). A tone and its octave are therefore very easily apprehended as one thing, and that unit of pattern may be followed with great ease throughout all sorts of changes of its volume as a whole. The doubling of a melody in octaves, then, is admissible in our music because it is quite easy to follow melodically and it is quite consistent with the volumic structure of our systems of intervals.

It has sometimes been said that the reason for the prohibition of consecutive octaves was that the effect of a four-part harmony was thereby lost. It is now evident from our system of facts and from the place of the octave in it that the reason cannot be of this merely negative order. The consecutive octaves must present a big positive something that is offensive. We shall form a clearer idea of what this is as we proceed.

CHAPTER XIV

EXCEPTIONS TO THE PROHIBITIONS OF CONSECUTIVES

In arranging the system of facts regarding consecutives we had to be content with an approximation to agreement in the statement of the rules. It was enough to bring out the general trend of the differences included within the system without striving to define it exactly in its absolute form. Fortunately there is not very much difference of opinion in the statements of these rules given in the chief text-books of harmony. In arranging the rules which state exceptions to the prohibitions of consecutives we shall again have to rely upon some estimate of the main trend. The general and growing agreement amongst theorists will facilitate our work for the present.

A full and sufficient account of these rules and their exceptions would best be based upon a very extended statistical treatment of the musical material. No doubt some of the theorists who have worked out rules of prohibition and of admission have collected large numbers of instances and have based their generalisations upon them. But for the fullest understanding much would be gained from an analytic study and a statistical manipulation of such a collection, if it were published in an extended form, so that the reader might follow the relative quantitative importance of the various factors that are found in groups of exceptions. What is required for an elementary knowledge of the principles of construction has doubtless already been attained. But even in some elementary matters these formulations have become detailed enough to show considerable divergence of opinion. This divergence is possibly not so much a sign of any difference in aesthetic reaction between persons or of any aesthetic differences in their nature, as rather a result of the consideration by each of them of different special groups of exceptions, or of only some of the factors operative in typical cases in abstraction from other accompanying ones that contribute to the final aesthetic effect. In any case there can be little doubt that an analytic study of a large number of exceptions on a statistical basis would be of great service both to the science and to the art of music. This sort of effort may be commended to those who have any favourable opportunity for making such large collections.

The exceptions admitted are few. For octaves (and fifths) Macfarren (35, 82f.) stated that the use of the sequence, "however rare, by composers of the present century, proves that this most stringently proscribed progression may produce an effect of measureless beauty, when it lies between the chord of the tonic and either 'dominant or subdominant,' provided only that, in the case of octaves, the parts that have the two in succession proceed by contrary motion" (cf. A. Day,. 10, 58). And Prout, as quoted above, agreed to this exception in the second edition of his work on harmony. Macfarren illustrates the point from Beethoven's Pastoral Symphony (between the bass and the alto) and from Sonata, Op. 53 (major common chord with doubled root in each hand on dominant and then on tonic). Similarly Tchaikovsky notes that in strict part-writing "(fifths and) octaves are permitted in the inner voices if contrary motion be employed" (76, 118). Parry (45) points out that consecutives are most objectionable in vocal and chamber music; in pianoforte and orchestral music they are often lost.

Shinn discusses the question at some length (58, 276f.), and claims that the sequence of octaves in contrary motion or of octave and unison often produces "an exceptionally strong musical effect." He generalises beyond the tonic-to-dominant or subdominant relation stated by Macfarren towards "other pairs of triads standing in a similar relationship with regard to their progression such as the triads upon the mediant and submediant," etc., "but these are not often employed." The musical strength of these progressions is not due "entirely to the fact of their harmonic relationship (or root progression), but partly also to the movement of the outside parts by almost equal intervals— that is, one by a fourth and the other by a fifth." The effect when one part moves a third and the other a sixth, especially when the bass moves the sixth, is generally less strong. Consecutives can rarely be employed in a satisfactory manner when one part moves a second and the other a seventh. It is not clear whether Shinn means the harmonic relationship of the chords or the mere movements of the voices to be the more important element in the effect; probably the latter.

Octaves in similar motion are admissible according to their purpose and position. As examples Shinn gives one (from Beethoven) for "the emphasising of a full cadence by the outside parts moving from dominant to tonic," another for the formation of a special melodic figure, and two between the final and initial chords of two sections. "In this position," he says, "their employment is by no means rare."

Shinn does not propose to sanction consecutive octaves when they occur in connexion with discords. Here we find further verification of the greater power of the octave towards bad effect.

As regards fifths, Prout's rule[1] may be taken as a generally accepted nucleus. His statement that the rule for fifths is much more frequently broken by great composers than the rule for octaves is well borne out by the relative frequency of examples to be found in text-books of harmony. From the various writers I have consulted (17, 23, 35, 38; 45, (48), 58, 61) I have collected some fifteen examples of octaves and over sixty examples of fifths. This relation may seem somewhat strange in view of the fact that it is customary to speak of consecutive octaves rather lightly and as being objectionable merely because of the temporary loss of distinction between the two voices[2]. But it is obvious that this theory was merely a deduction from the notion of the musical equivalence of octaves; it did not properly reflect the nature of the musical phenomenon itself. And some writers even proceed to explain the bad effect of fifths by the loss of independence of the voices they appear in.

As regards the influence of the progression referred to by Prout, Gladstone wrote that "of the various exceptions which the great composers have made to their rule of avoiding fifths, none are more common than those in which the progression is either from the tonic to the dominant, from the tonic to the subdominant, or the reverse of either" (17, 103f.). This is the first stage of his argument in favour of explaining the effect of fifths by the relative position of the chords in which they occur. It is again to be regretted that his statements were not accompanied by some evidence showing relative frequencies. For the next degree of relationship (a third between the roots) he could only cite two cases with an ascent of a sixth between the roots and none with a third.

In connexion with fifths Shinn (58, 280 ff.) seems to rely entirely upon harmonic relationship, making no allusion to the movements of the voices. He indicates a decreasing frequency of occurrence and a loss of

---

[1] Of contrary motion A. Day wrote: "Fifths by contrary motion should not be used (although by most writers allowed), as the reason given why fifths by similar motion should not be used [they give the idea of two different keys] is equally applicable to fifths by contrary motion" (10, 10). So much the worse for the reason given, one might rather say!

[2] Pearsall, for instance, dismisses consecutive octaves and unisons in five lines (of his 27 page quarto pamphlet), saying they ought to be avoided because of 'awkwardness,' and "because they produce no effect except that of rendering insipid and almost nullifying any harmony of which they may be component parts" (49, 26).

effect in the series of relationships of a fifth or fourth, a third or sixth, and a second or seventh. In the last case the effect is rarely satisfactory— except for special purposes—in the root position of the chords; it is better in inversions. Shinn gives many examples of sequences of fifths in connexion with *discords*[1] (essential and unessential) and suspensions. The explanation he offers of these is that the dissonant note imparts to the chords "such a new, distinctive, and relatively speaking, forcible character, that the unpleasantness due to the consecutive fourths [fifths], if it is not obliterated, is neutralised by the introduction of this new element and the sound of the progressions becomes entirely satisfactory. Not only is this explanation perfectly adequate, but it is, we believe, the only one which it is possible to supply that is based upon the musical effect of such progressions" (58, 286 f.).

This last exception claimed by Gladstone and Shinn is confirmed by the large number of instances of it that are to be found in writers on harmony. I have collected and compared more than sixty cases of consecutive fifths, mostly from the greatest and most accepted composers[2]. Of these 11 show a progression from a discord (commonly a minor seventh) to a concord, 16 from a discord to a discord, and 15 from a concord to a discord. In reckoning these numbers I have counted as one only one type of progression in a given work. Thus a seventh to a seventh would reckon as one, no matter how often it were repeated; but a seventh to a common chord ending the passage would count as a new case. Only five of the 42 cases are by contrary motion. Sixteen lie between the bass and the soprano (B–S), 10 are B–T, 8 are S–A, 7 are A–T, and only one is S–T. In resolving these sevenths naturally pass sometimes to dominant and to tonic; but this is a feature of the progression that must be considered accidental; it is not what legitimates the succession of fifths.

That can only be the discord in question. And the effect produced is a reasonable one[3]. In this type of exception the two ends of the

---

[1] Day held that fifths by contrary motion are allowed if either of the chords or both be one of the fundamental sevenths (10, 59).

[2] The examples collected by Parry (48, 119 f.) that he dubs with the scornful title "music-hall cadence" do not all deserve so severe condemnation when considered in this connexion, whatever other faults they may exemplify.

[3] Sacchi said consecutive fifths were admitted "(1) When the fifths lie between the inner parts, not between the outer ones, (2) when the fifth with the bass is covered by the sixth, (3) when the two fifths are so placed that the first ends one period, while the second forms the beginning of another." Sacchi's explanation of his second exception is excellent: "The fifth being here covered by the sixth, this consonance cannot be clearly

fusional series are put into operation against one another, both being intervals characteristic of chords—not a sort of repeated note, as the octave may often be. The consonance makes too much (perpendicular) unity or fusion, the dissonance too much (perpendicular) duality or ruption for proper melodic flow. But their combination gives a new balance of flow. Only,—and that may well be a notable practical point, —the combination is most frequently such an interlocking of the two elements as prevents either from standing forth and dominating the progression.

This is exemplified in my sample. In 16 cases the fifth lay B–S; of these only two are by contrary motion. The beneficial effect of contrary motion is evidently not required for B–S. In 7 cases the position of the fifths is A–T, and all of these are by similar motion. Ten cases lie between the bass and the tenor or the voice next above the bass. In this distribution we should expect the fifths to be more liable to fall away from the rest of the chord, and so to become more than usually noticeable. This view is perhaps supported by the fact that in three of these cases contrary motion has been used, that in two the harmony is of six parts, that one is produced merely by a sort of shake in the tenor, and that in two one of the voices is helped to continuity by an inserted passing note. Thus where there is special danger of the inter-locking of the two opposed elements in the chords being lost, there the composers have brought other compensatory influences to bear upon the fifths. In one case S–A with similar motion the progression is to a chord of the minor seventh, but support is given by special melodic features, "the carrying out of a thoroughly established idea," as Parry (45) says of this example. In one by Chopin the fifths occur as a tremolo-like accompaniment to the bass figure. Five others are by Dvořák, and one by Stainer.

Amongst the cases involving no discords the fifths are referable in

enough distinguished. The sixth is here much more noticeable than the fifth, because it is the extreme parts that most strike the ear and draw the greatest attention. Besides, the two neighbouring notes, the fifth and the sixth, form a dissonance, a so-called acciacatura; and the effect of the acciacatura is a certain suspension and indecision of sound, that makes us expect its resolution."

The admirable Sacchi, in fact, ends on an ultra-modern note by saying that we must exercise moderation in judging of consecutive fifths, and that we must always consider, besides, "the beauty and novelty of the thoughts, the regularity and artistry of the progressions, the elegance and clearness of the melody, the unity of the design, the force of the expression," and the 'convenevolezza del costume' (the propriety of the feeling?). In short we have to be equipped, not only with "the eyes of the face, but with those of the mind" (57, 81 ﬀ., 86 f.). Plenty of scope for freedom and progress there!

ten cases to the use of passing, or more or less purely ornamental, notes. They are all by similar motion.

The remainder include two by Mozart, one by Mendelssohn, one by Rheinberger, the famous one from Beethoven's Pastoral Symphony, three by Schumann (one between the beginning and end of phrases) and one by Elert. Four of the preceding are by contrary motion. Two (Prout and Gounod) are evidently intended to give the effect of barbarous progression.

Among the non-discordant cases that by Karg Elert (23, 10) is not only very beautiful, but at the same time unique in its build. It would seem as if the two series of neutral intervals (sixths and tenths) were able to outweigh that of the fifths:

KARG-ELERT'S EXAMPLE

Karg-Elert, " Näher mein Gott "

Sw. Salicional.

While progressions (involving fifths) between tonic and dominant or subdominant of course occur, many other connexions appear in the most pleasing cases. From the cases I have collected I cannot persuade myself that the harmonic connexion of chords (with the probable exception of the one just mentioned) plays the important part in the degree of acceptability of consecutives that Gladstone and Shinn suggest. But I do not wish to put forward the groupings I have just given as more than mere suggestions. The subject calls for an extensive treatment on statistical (and experimental) lines, and for that the small number of examples I have collected is only profitable in broadest outline. In all probability they form a very special and biassed example. But there is no doubt that the effect of the discords on the fifths stands out prominently even so. And the features of the pleasing uses of fifths they reveal, generally seem to be compatible with the theory of the basis of the prohibition that was advanced in the preceding chapter. They all either obliterate the special effect of fifths or they go to strengthen what the fifths weaken—the even flow of the different voices.

For consecutive fourths with the bass it is stated by Shinn (58, 166) that the prohibition does not apply to progressions in which either of the chords forming the progression is a discord. Prout, as we saw, allowed them only "when the second of the two is a part of a fundamental discord or a passing note" (cf. 58, 285f.).

CHAPTER XV

HIDDEN OCTAVES AND FIFTHS, ETC.

In all treatises on harmony a regular counterpart to consecutive octaves and fifths is found in the special treatment of so-called hidden octaves and fifths. The traditional explanation of the latter sets them into direct connexion with the former. Two voices that approach a consonance by similar motion are supposed thereby virtually to present consecutives. Only, as the tones leading to the consonance are not really sounded, they were styled 'hidden' (cf. 55). The theory is ingenious in so far as it thus leaves only one thing to be explained,—the bad effect of consecutives. But Shinn (59) is not far wrong in saying that this theory "can only be regarded as an interesting tradition of the past, which, in the present day no intelligent musician can pretend to believe." We are the more relieved from discussing it, as it rests not on any facts, but on a mere assumption, namely that the listener *unconsciously* fills out these slurs. Some objective strength might be given to the position by claiming that there is some sort of a real melodic slur involved in the process; but even that theory would hardly get beyond a preliminary formulation. For this melodic slur would not bring the phenomenon into relation with consecutives for which the presence or activity of a melodic slur has never been assumed.

As we have learned in the preceding, our best method will be to find and to describe the system of facts of which 'hidden' consecutives form a part and to try to draw from this system an interpretation sufficient to describe, and in describing to explain, its members and their relations. And as the objective facts summarised by exponents of systematic harmony themselves suggest the procedure, we may take our systematic table of consecutives as a model or ideal of discovery in dealing with 'hidden' intervals. As before we shall set out from the formulations of E. Prout (52), which have been summarised in the previous manner in Table III.

TABLE III

' Exposed ' Intervals (52 28 ff.)

Showing relations between (1) the stringency of prohibition, (2) the grade of fusion of any interval, and (3) the prominence of the voice-parts in which the interval appears. 'Exposed' intervals (or 'hidden' octaves, fifths, etc.) are such as are approached by 'similar' motion in the two parts, i.e. both parts rise or fall (in pitch) to the interval in question.

	O.	5.	4.	Third or sixth	T	Seventh	Second or ninth
B–S	forb.	forb.	+	+	?	?	?
B–A	f.	+	+	+	+	+	+
B–T	f.	+	+	+	+	+	+
S–A	f.	+	+	+	+	+	+
S–T	f.	+	+	+	+	+	+
A–T	f.	+	+	+	+	+	+

forb. = forbidden with exceptions; f. = hardly forbidden; + = allowed; ? = recommendation against.

The formulations embodied in the Table are not held to be exact laws, as it were, but only to be properly representative of the *trend* of opinion on the subject *in its grading of stringency* from point to point of the Table.

The rules upon which this Table is based are the following (52, 28 ff.) :

(1) Hidden octaves are forbidden between the extreme parts; except, first, between primary chords in root-positions—(i.e. with the roots in the bass), when the bass must rise a fourth or fall a fifth, and the upper part must move by step; second, when the second of the two chords is a second inversion[1], the bass note being either the tonic or dominant of the key; and 3rd, when the second chord is another position of the first.

Hidden octaves are, however, allowed between any other of the parts than the two extreme parts, with one important exception. It is strictly forbidden to move from a seventh or ninth to an octave by similar motion between any two parts, when one part moves a second, and the other a third. This is the very worst kind of hidden octaves, and must be most carefully avoided.

(2) Hidden fifths are forbidden between extreme parts; except, first, in a progression between primary chords (tonic to dominant, or subdominant to tonic), with the upper part, as with hidden octaves, moving by step. The first of the two chords is not (as in the case of octaves) restricted to root-position (two examples are given of it in first inversion); second, from the root position of the chord of the supertonic, with the third in the upper part, to the chord of the dominant, when the bass falls a fifth, and the upper part falls a third; and third, from one to another position of the same chord, exactly as with hidden octaves.

---

[1] Here there is sharp opposition with Tchaikovsky who says (76, 60): "Concealed octaves sound particularly unpleasant when the octave appears as doubled Fifth in the six-four chord (or as doubled third in the sixth)." In both examples given by Prout as good the octaves are given by a doubled fifth. This disagreement between the two theorists is not due to confusion of the requirements of the hidden octave with those of the fourth from the bass, but reflects their attitude towards the hidden octave only.

(3) When two notes making a dissonance with one another (such as second, seventh, or ninth) are taken without preparation—that is, if neither of them has been sounded in the same voice in the preceding chord—it is better that they should enter by contrary than by similar motion, especially in the extreme parts.

Tchaikovsky (76, 59 ff.) has treated hidden octaves and fifths in some detail. Of the examples he gives as bad all would be forbidden by Prout, except the progression to a six-four chord with doubled fifth cited above; Tchaikovsky, besides, allows the octaves or fifths "when they occur in the connexion of a triad with the Dominant chord, and when the seventh is prepared in an inner voice," a case that Prout does not mention. Tchaikovsky's classifications overlap one another considerably, so that it is difficult to see clearly how far they coincide logically with Prout's. They may be summarised as follows :

(1) Octaves and fifths in the outer voices are often very disagreeable and should be avoided by all means : (a) when the upper voice proceeds by a jump, but even then the progression is by no means unpleasant in the connexion of a triad with the dominant chord when the seventh is prepared in an inner voice; (b) when all the voices move in parallel motion; (c) when the outer voices move in parallel jumps, even though the inner voices remain stationary or progress stepwise.

Here there is evidently an opposition between jumps and steps or stationariness : (a) one jump in the upper voice, (b) one jump worsened by more parallel motion, (c) two jumps bettered by steps or stationariness. Prout refers to step only in the upper voice, seeming thus to indicate greater importance in the melodic continuity of that voice (even over the bass)[1].

(2) Concealed progressions between an outer and an inner voice are disagreeable when the jump lies in an outer voice. (C. H. Kitson—30, 50—formulates a similar rule for three parts as against four parts, for which it is not required.)

(3) Between the inner voices concealed octaves are "entirely out of the question, arising solely in consequence of bad voice-leading"; "concealed fifths, however, are permitted, provided the voice leading is natural." (The word 'however' seems to imply that the octaves are here meant to be more forbidden than the fifths.)

On the whole we may infer a grading of prohibitions, greatest for B–S, least for A–T and medium for the intervening pairs (cf. 55), while the minimum for the fifth in A–T is less than for the octave. Thus we get from Tchaikovsky's analysis the same general result as from

---

[1] But cf. below, p. 130, note 2.

Prout's. Of the examples given by Prout as good or bad all would be similarly styled by Tchaikovsky except (1) the six-four case already noted, (2) the progression to another position of the same chord, which Tchaikovsky neither mentions nor exemplifies, and Prout's second exception to the rule against hidden fifths, which stands in systematic isolation amongst Prout's other rules, as far as one can see.

Jadassohn (26, 36, 70) likewise forbids concealed octaves and fifths when both voices leap, "no matter in what connexions and in what direction, if in outer or inner voices, or one outer and one inner voice," except when it is a case merely of inversion, or of interchange of voices. One leap is not so bad, especially if it is not in the upper voice (Sopr.), even in a progression between the triad on the mediant to that on the submediant, or similarly from the submediant to the supertonic (26, 37). There is some sign of greater freedom with the fifth than with the octave : for "if the upper voice moves by a degree, and one of the lower voices by a skip, the concealed fifth between all the voices is without hesitation permitted, provided all the voices do not move in the same direction[1]" (26, 72); in the case of octaves a slight reservation seems to exist in the case of connecting the chords on the second and fifth degrees (26, 37). In the chord of the dominant seventh a fifth may occur even after similar motion of all the voices, but not along with a diminished fifth.

The exceptional treatment of the progression from a chord on the second to one on the fifth degree may have some connexion with Prout's second exception to the rule for hidden fifths, but it seems to contradict it rather than to support it. The systematic connexions of this case are not at all evident. Jadassohn also cites the modifying effect of pitch-direction on this succession : "Concealed octaves in the succession of the chords on the second and fifth degrees, when downward, are not faulty, as this connexion does not sound harsh" (26, 35); they must, however, be avoided in upward progression (26, 36). These points do not affect the main result, but only point to other modifying factors, which would each require a separate treatment, i.e. a search for their systematic setting.

One other theorist is worthy of mention. Shinn (59) says :

Viewed from the modern standpoint, the term 'hidden' itself—seems a misnomer.... When so-called hidden octaves and fifths do produce an objectionable musical effect, it is obviously due to the fact that the octave or the fifth which *is* present is itself thrown into undue prominence, and what is unsatisfactory in

[1] Cf. below, p. 126.

the progression is the result of the exceptional 'exposure' of the perfect interval which is present, and not to a faulty progression which is supposed to exist in the imagination of the listener. This fact is now admitted by the more progressive amongst musical theorists, and in modern text-books the traditional and misleading term 'hidden' is being gradually displaced by the more accurate one 'exposed,' the universal adoption of which is certainly desirable in the interests of all students and teachers of Harmony (cf. 58, 265 ft.).

It is disappointing to find so interesting a statement made without any citation of the author of this change of interpretation. I have not as yet met with the term 'exposed' in any other writer on harmony except C. H. Kitson (30, 49) and prefer meanwhile to consider F. G. Shinn himself to be the author of any distinct theory of 'exposure' in these intervals, as he is certainly the only satisfactory exponent of the idea. It is obvious that his interpretation is much more in line with the trend of our own exposition than is the older term 'hidden,' although the latter originated quite properly in an attempt to connect this set of facts with that regarding consecutives. The mistake the traditional theory made was to cling to the mere external form presented by the consecutives as the essence of their offensiveness instead of getting behind that form to the true essence. That essence, we have seen, is most probably the break in the usual melodic connexion due to the greatness of the consonance or dissonance in question and to its 'exposure' by similar motion.

Shinn then points out that "some of the most eminent continental theorists impose restrictions upon the employment of octaves and fifths so formed [exposed], either between two inner parts or between one inner and one outside part, which are not recognised by English theorists." He then cites Tchaikovsky and Jadassohn and remarks in connexion with an example constructed to fit the rule quoted from Jadassohn above (p. 125, note):

When such restrictions and prohibitions extend so far as to describe as bad and to forbid the employment of ["such a progression,"] it is doubtful whether such rules are not merely devoid of all musical authority, but whether they possess any value even for the purposes of mental discipline; whether, in fact, they do not tend to make the introduction of any kind of spontaneous musical thought into the work of the student absolutely impossible.

He himself then proceeds to consider their employment "from the point of view adopted by English theorists, that is, when they are formed between the outside parts"; and he suggests on the basis of

the examination of many examples of exposed octaves and exposed fifths...that the effect of such progressions varies (and is more or less satisfactory) according to the extent to which the particular effect of the exposed interval dominates the effect of the second chord of the progression. When the nature of the individual chords forming the progression, their harmonic relationship, the number of parts employed, or their general progression, is such as to neutralise the effect of the exposed interval, no unsatisfactory effect is produced.

By the nature of the individual chords is meant whether either or both of the chords be a discord, or whether both are concords. When the second chord is a discord, to whatever extent an exposed interval may be thrown into prominence, the effect of this exposure is almost invariably neutralised by the dissonant character of the chord.

Shinn gives two examples, one for the octave and the other for the fifth, each occurring in a chord of the minor seventh (on supertonic and on dominant).

He then discusses the effect of harmonic connexion upon 'exposure' :

In connexion with triads and their inversions whose roots are a fourth or a fifth apart, and which therefore have one note in common, when one of the parts moves by step, the other part leaping, the effect is rarely unsatisfactory. When both parts leap, especially in a downward direction, neither part leaping more than a fifth, the effect may be excellent.

When the roots of chords are a third or sixth apart, the effect depends partly upon the degrees of the scale upon which the chords stand, "some being bold and strong, others weak and unsatisfactory." Downward approach is usually a favourable circumstance. The degrees used also affect triads on adjacent notes. Inversion also :

When one of the chords is an inverted form, and the highest part moves by step, while the lower part leaps either a fourth or a fifth, the effect is almost invariably good.... An exposed octave formed between two such triads, the second being in its second inversion, is also unobjectionable when one part moves by step and the other leaps a fourth.... The strong and characteristic movement of the two parts, by a second and by a fourth or fifth, exerts considerable influence in the direction of strengthening the effect of the progression (cf. 58, 272 ff.).

Here we see a number of factors influencing the essential factor in question—the degree of consonance of the ('exposed') interval; we can see to some extent the direction of the influence, whether it is favourable or unfavourable and we can perhaps assess them against one another. But the general result is hardly as clear as is desirable for scientific aesthetic purposes. Finally, he points out that "progressions which are unsatisfactory in two or even three parts may be quite good in four or five parts." This he says is due partly to the doubling of notes common to both chords, partly to the distribution of the listener's attention over a greater number, when "the progression of any two parts (even when they are outside parts) must in some corresponding proportion become less noticeable" (cf. 50, 255 f., 296).

The result of this analysis of representative authors may seem hopelessly confused, as it must certainly be highly unsatisfactory from any practical point of view[1]. But we can skim the common general effect off the differences peculiar to the different writers. (1) The objectionable phenomenon in question is presented by the octave and the fifth (all theorists agree), and somewhat more strongly by the octave than by the fifth (Prout, Tchaikovsky, Jadassohn and Shinn)[2]. But whereas in the case of consecutives the objectionable feature appeared regularly and was only mitigated or made tolerable by contrary motion, in this case (2) it does not appear until it is 'exposed' by the unfavourable effect of similar motion. Parallel motion in other voices than those primarily concerned increases the unfavourable effect (T., J.). (3) The effect is worst in the outer voices (P., T., S.), it is least in the inner voices (T.) and of medium degree between an outer and an inner voice (T.).

The unconcern of the English exponents to any but the outer voices is, of course, not inconsistent with a gradation of the acceptable effect in the rest of the series of voice pairs. Differences of opinion on the position of the point of change from desirable to undesirable effect are inevitable in the midst of so many fluctuating factors. It would be a mistake to wish for that reason to throw aside all the formulations of previous exponents, even if we feel that their terms do not properly express our judgments, and to start a search for an entirely new set of formulative notions. A prejudice of that kind may make success impossible. We should keep our attention chiefly directed upon the systematic trend of the rules of previous analysts, in case our difference from them may simply be settled by a shifting of the border line of acceptability without any radical change of the basis of judgment.

This counsel holds not only for the primary features of the situation, but also for the minor factors which create exceptions from the main rules.

Thus (4) the progression by step has a favourable effect, while leaps

---

[1] Kitson says (30, 49): "It is impossible to find any basis of general agreement as to which exposed consecutives are objectionable and which not."

[2] Shinn says "the effect of exposed octaves distinctly differs from, and is often far less satisfactory than that of exposed fifths" (58, 271).

are unfavourable. Two jumps—one in each voice—are worse than one, while the effect of any one jump varies with the prominence, especially the melodic prominence, of the voice that bears it. A step accordingly is more powerful in the highest voice[1], then in the lowest, etc. Jadassohn says jumps in any two voices are bad, while one is least bad if it is not in the upper voice. Tchaikovsky deems two jumps in the outer voices very bad, one jump worse in an outer than in an inner voice, and not usually admissible in the highest voice. Prout requires a step in the upper voice to support the favourable influence of close relationship of chords. And Shinn thinks the strongest effect is produced when the upper part moves by step (58, 272 ff.)[2]. He also desires two jumps (in outer voices, of course), to be mitigated by downward motion and restriction of the leap to the 'emmelic' range (not more than a fifth).

Melodic continuity, then, counteracts the bad effect of the consonance exposed by similar motion. This agrees with our conclusion for consecutives, that their offence was a breach of the required melodic continuity. (Two steps would, of course, often convert the problem into one of consecutives. A tritone may be followed by a perfect fifth if the lower part rises a semitone; then there are two 'steps.' Cf. p. 104 above, Rule II.) But while there is thus agreement in the general trend, there is considerable disagreement between writers as to the margin of pleasant effect. But the drawing of these border lines does not primarily concern us here.

(5) Writers (T., J., S.) agree that the bad effect is covered over again in characteristic discords, e.g. that of the minor seventh. One might argue that the exposure should then be greater as the two intervals stand at the ends of the fusional scale. But the direction of pull, as it were, is opposite in the two cases: one is consonance or unity, the other is dissonance or duality. Perhaps this combination achieves something like the neutrality of the thirds and sixths (without, of course, rendering the chord consonant as a whole).

(6) 'Connexions' between the chords or the tones seem favourable. When the chords only differ in position, the bad effect is annulled, probably because melodic connexion is readily attained from one

---

[1] Which in modern music is usually not only generally, but also thematically, melodic. Cf. Tovey's definition of melody as "the surface of music."

[2] In the case of the tritone followed by the fifth, already referred to (p. 104), the progression is satisfactory in so far as the leading tone, which forms the lower note of the tritone, moves by semitone to the tonic. But if this movement is exposed on the bass, it so strongly 'announces' the coming fifth that we get a specially exposed fifth, or else the leading tone refuses its chief function by falling (cf. 52, 27, 101).

position to another of the same chord or pattern. Here the second tonal mass has been facilitated for the mind by the occurrence of the first one, so similar to it. Thus the melodic connexions are favoured. Similarly when the bass moves from one important, and therefore familiar (or easy), point of the tonal system to another (tonic to dominant and *vice versa*, subdominant to tonic and *vice versa*), or when it merely moves by similar intervals, fourth and fifth), continuity in the bass is made easy; and if step relieves any difficulty in the upper voice, the effect should be excellent[1]. But there seems no strong reason why with Shinn these relations should not be inverted to step in bass and easy leap (e.g. fourth or fifth in upper voice). Only, the system of gradations we have shown would incline us to expect not quite so good an effect in the latter way as in the former[2]. But, of course, both ways may be acceptable. The degrees of the scale used, other than tonic, dominant, and subdominant, may further modify the ease of progression according to their familiarity in the tonal system.

(7) Prout's note regarding the seventh or ninth before the octave seems clear, because these intervals, even in isolation, strongly suggest the unity and repose of the octave. This suggestion is probably even increased by the stepwise movement of one of the voices, which will ordinarily be accompanied by the movement of a third in the other voice, when the following interval is an octave. Thus the octave about to be heard will be strongly suggested, and so will be more exposed than ever.

(8) In dealing with consecutives we have already encountered the effect of a larger number of parts than four. Prout (52, 305) says of this: "In proportion as the number of parts, and therefore the difficulty, increases, the stringency of the rules relaxes. These hidden fifths and

---

[1] These points are nearly all applicable to the unison, which may not generally be taken by similar motion, unless in the progression from dominant to tonic, with the help of step in one voice also perhaps (cf. 52, 31 f.). The voices in which the unison appears will generally be neighbouring ones, of course. The unison by similar motion seems to be more strictly forbidden than is even the octave. The unison, of course, gives the clearest expression to the characteristic of consonances—their apparent unitariness.

[2] This would imply that the leap of a fourth or fifth is a more potent factor than is a step. For, as the bass gives greater exposure than the soprano, x (fourth or fifth in bass) $+y$ (step in soprano) would be greater than y (step in bass) $+x$. This inference might be preferable to that suggested on page 124 above (greater melodic prominence of the soprano), which is irreconcilable with the general predominance of the bass except in so far as the soprano in modern music is often thematically the most melodic voice, though it can never be generally—or in its mere sonorousness—the most melodic. Cf. the later discussion on the most general aspect of melody.

octaves are allowed, even when both voices leap; consecutive octaves and fifths by contrary motion may be used freely; we even meet in the works of the great masters with examples of a doubled leading note, though it is better to avoid this, if possible." It would probably be wrong to suppose that the mere difficulty of the work is the basis of license, as if an ugliness could be compensated by such an extraneous reason. It seems preferable to suppose that the succession is admitted because the many parts steady each other and maintain a general distribution of effects. The obscurity of the inner parts in four part writing has the same origin.

We shall have some opportunity later of considering the influence of inversions. And the difference between upward and downward motion may be neglected for the present; it is not quite clear what the relation of this difference is to the matters referred to in chapter VIII. The implication is that descent strengthens melodic continuity, possibly because descent has more of the character of a return to the starting-point, while ascent is departure (cf. above, p. 52 f.). Prout's unique rule regarding the supertonic chord is probably more or less of an accident in his usually so systematic work. It seems a composite of downward motion and of progression of a fifth in bass. Only the latter factor is invoked by Shinn to explain the case (58, 49).

In conclusion we may say that the Table at the beginning of this chapter, based upon Prout's formulations, properly represents the *trend* of differences it refers to, and may therefore be taken as valid material upon which a theory of the phenomenon in question may be raised. As we have so often suggested, the exact place at which one draws the line between what is desirable and what not, depends on variable subjective factors. But the scale of differences which make different aesthetic reactions possible is objective and invariable, as is also the general trend of their influence upon the aesthetic reactions of an individual. The aesthetic realm is not to be considered the sport of caprice because there is no disputing about tastes. The latter statement is true only because one man may carry subjective inclinations and influences about with him that another does not possess and finds of no particular interest. If these so affect him as to make his aesthetic judgments contrary to another's, the latter cannot offer to dispute him out of their influence. Between one man and another only those forces are subject to common analysis or discussion that are objective to both of them, or that are primarily rooted in the artistic work itself.

A study of the factors that operate in some individuals and not in others is work for a psychology of personal differences, not for a science of the constitution and potencies of the aesthetic objects themselves. About these and the laws of their being dispute is as little excluded as it is in dealing with nature itself.

It is often said that artistic rules are hardly formed before they are swept away into oblivion by the stroke of some genius. That and all such expressions are radically wrong, as wrong as it would be to say that no sooner is a law of nature discovered than some engineer sweeps it away by showing how to circumvent it or (apparently) to oppose and to reverse its action. The will of a person may be overruled and forgotten, but no one supposes the art of a past century to have been a tyrant's will. A genius breaks no rule of art. He only fulfils it the more by finding influences which unite with it to produce effects it would be incapable of producing alone. After all no one really believes this fable of the genius. You always have to be the genius before you can have his power to make rules disappear. You must have his knowledge and experience. In fact, you must know[1] how to do it. And it is right that the beginner should learn the big facts and rules first, as really and permanently valid, not irreconcilably valid or as pedants' foibles to be discarded later. He must first learn the broad effects and then progress towards the subtle ones; and he will do so easily and willingly when his discipline can be set before him in systematic form. When its foundations have been well expounded, he will be able of himself to carry them forward into their interactions with one another much more readily and steadily than he would if he had to learn them all separately and unintelligently.

---

[1] Or 'feel' (in sensory constructions). A composer who does so, may perhaps not also *know* the method or law inherent in his feeling. It is for the theorist to find that out, if the composer does not first discover it himself. In his " Philosophy of Modernism (in its connexion with music) " Cyril Scott says: " It is a fact, almost a truism, among enlightened musicians, that we learn the rules only in order to know how to break them; but the real quarrel arises in how often and how far one is permitted to break them. The truth is, *in reality there are no rules*. There are merely *conventions*; and these conventions have altered with the advent of each new master " (p. 27 f.). ' Breaking them ' suggests the overcoming of law by its better fulfilment; but, of course, there is also such a thing as the mere ignoring of rules.

CHAPTER XVI

A FOURTH FROM THE BASS

It is a notable fact, which was emphasised in the preceding chapter, that similar motion is the primary condition of the prohibited 'exposure' of the two intervals octave and fifth. The next most important condition, embodied in the Table on page 123, is the prominence or exposure of the pair of voices that bear the interval. We noticed that amongst English theorists the prohibition of exposure extends only to the most prominent pair—bass and soprano. The intervals prohibited are familiarly octave and fifth only; but it is proper to place the dissonances at the other extreme of the same Table because of the common desire that they should not be exposed by similar motion.

We have thus detected at least a part of a system of facts that forms a proper counterpart to the system obtained for consecutives. We might offer to pass the rest of the system as the counterpart of successive thirds and sixths—the neutral region for which no prohibition exists. The only difference would be that this neutral region is now wider. And that would be in natural agreement with the greatly reduced degree of prohibition set upon 'exposed' intervals. We might even infer from the scope of this widening that the fourth as a consonance borders more closely upon the neutral region than its common ranking as a perfect consonance would suggest. That would also agree with the system of consecutives; for in it we saw that only fourths from the bass are forbidden. Or we might say that the fourth lies on the border between high grade consonance and neutral sonance.

But we cannot accept any such simple solution without careful inquiry. For it is a familiar fact of harmony that a fourth from the bass must be used with great care. Many rules for its use have been formulated and these must be carefully analysed before we can judge accurately of the status of the fourth as an 'exposible' interval.

We shall again base our summary in the first instance upon the formulations of E. Prout (52, 66 ff.).

"Though it is possible," Prout says, "to take any triad in its second inversion, the employment of any but primary triads in this position is extremely rare." Macfarren (35, 68 ff.) had asserted that only those on the tonic, subdominant, and dominant were admitted. It is, of course, the 'root' of the chord that is supposed to stand on these degrees; the lowest tone of the second inversion is a fourth lower. For purposes of description these two authorities may be said to agree. The six-four chord has not only to be approached, but also to be quitted, in certain ways.

It may be approached : (1) either by leap or by step from the root position of another chord; (2) by leap from another position of the same chord; (3) by step (but not by leap) from the inversion of another chord; (4) from a different chord upon the same bass note. These steps and leaps all refer to the bass note upon which the fourth stands.

It may be left : (1) by step of a tone or a semitone, upwards or downwards, and the following chord may be either in root position or in an inversion; (2) by the same bass note or its octave bearing another chord, provided the six-four chord is a tonic or a subdominant chord. (This is the 'cadential six-four.') And in this case the six-four must be on a strong accent, unless it has also been preceded by a chord on the same bass note. (Evidently this cadential effect requires a sort of rhythmical exposure of the six-four chord to create a ready disposition for it in the listener's mind.) (3) By leap to another note of the same chord (without change of harmony), provided that, when the harmony changes, it returns either to its former note, or to the note next above it or below it. In other cases it is not good for it to leap (32, 70f.).

Now we know already that for melodic continuity a leap is less favourable than a step, and the repetition of the same note is easier to follow than is a step. By abstracting this term from the above rules and by grading those that are left over against it, we may attempt to gauge the effect of the latter. By this process Table IV has been constructed. In order to enter a single value at each point I adopted as indicator the *extreme* permitted under each head. Thus 'by leap or by step' means 'even by leap'; if a leap is permitted, so is a step, of course. Similarly 'inversion' means 'either inversion or a more favourable form—root position.' But 'root position' means 'only root, not inversion.'

The Table, then, indicates that the melodic continuity and ease of the bass voice in a six-four chord are a matter of special concern. For when favourable conditions are secured for it by conjunct motion or by no change at all, the other parts have complete freedom of movement; but when its movement is made less cogent by a leap, the others should be more favourable, giving at least either the same chord or (in approaching the six-four) proceeding from the clear cut stability of a root position.

TABLE IV

Scheme of rules (after E. Prout, 52, 70f.) for approaching and quitting a six-four chord.

	Approaching a $\frac{6}{4}$		Quitting a $\frac{6}{4}$	
Movement of the Bass	From— $\left(\uparrow\right)$ chord	In— $\left(\uparrow\right)$ position	To— $\left(\uparrow\right)$ chord	In— position
By leap	Same $\}$ $\uparrow$ Different $\}$ $\uparrow$	$\downarrow$ $\{$ Inversion $\{$ root	Same (1)	Inversion
By step	Different	Inversion	Different (3)	Inversion
None (or octave)	Different	Inversion	Different (2)	Inversion

The arrow-heads point in the direction of greater melodic continuity and ease. Each entry in the Table indicates the *maximum* allowed.

Note 1. Provided that when the harmony does change, it returns either to its former note or to the note next above it or below it (or to a note to which a correct progression from the six-four chord could have been made, 58, 61).

Note 2. Provided the six-four chord is a tonic or a dominant chord. And in this case the six-four must be on a strong accent, unless it has also been preceded by a chord on the same bass note (cadential six-four).

Note 3. In this case the six-four should occur on the unaccented part of the measure (76, 55 f.). If the one is followed by another six-four, the latter will be cadential, and so a rule of frequency for successive six-fours may be formulated (cf. 38, 50).

The rules given by Prout thus evidently form a consistent whole, which is at the same time thoroughly representative of the generalisations of other writers on harmony. Minds may, of course, as we have already noted, vary for subjective reasons in the exact margin of desirability they draw. For a study of the effect of objective differences that concerns us now, the essential consideration is the trend of the changes in desirability. The smaller the range of personal differences, the more reliable will this trend be. And here the range seems to be very small indeed.

There is, then, in the six-four chord a sort of opposition or rivalry between the two parts of the chord,—the bass and the other parts. The problem is : what is the basis and origin of this melodic rivalry? A strong indication is given by the fact that the essential characteristic of the chord is its fourth from the bass. This has been condensed into the radical classification of the fourth as a dissonance. Such a theory is, of course, erroneously extreme : the fourth cannot be set down generically as a dissonance, for apart from the bass it is not at all dissonant. But the indication at least requires us to consider whether any special treatment has to be given to the upper note of the fourth.

Prout makes no statement on this point whatever, beyond pointing out that the upper is the dissonant note and therefore should not be doubled (52, 71). Macfarren (35, 90) pointed out that "the consonance of the fourth in these three inversions [on tonic, subdominant, and dominant], is proved by the entirely free progression of the 4th it comprises, which is the assumed dissonant note of the disputants." Shinn (58, 61) says the rules for six-fours "refer entirely to the progression of the bass part." On the contrary Tchaikovsky says that the six-four chord "must in any [other?] case be connected with one of the neighbouring chords by means of its fourth; and into and from its other neighbour the fourth must progress stepwise" (76, 56). Mansfield (38, 38) is more comprehensive : "The 4th ... being a dissonant interval should be approached and quitted conjunctly, and, if possible, in contrary motion with the bass. Failing this, it should be prepared, i.e. heard as a consonant note (or as an essential note, i.e. a note without which a chord would be incomplete, such as a 7th in the chords of the dominant and diminished 7ths) in the same part in the previous chord. If approached by skip, it should be in contrary motion with the bass, and it should be quitted by oblique motion with the bass when contrary motion is not possible. These rules apply to almost all dissonant notes." Table IV shows that oblique movement is frequently implied in the rules, e.g. over against 'leap' and 'none.' Opportunities for similar motion would arise mainly where the bass moved by step. Whether the number of times it actually occurs is so small as to make a prohibitory rule useful I do not know. One of Shinn's examples (58, 60, e) shows similar motion.

In favour of the consonance of the fourth there are to be urged— the work above expounded regarding fusion, a large part of all musical experiences relating to the fourth (i.e. all apart from the bass), its graded position in the Table of consecutives, and the beneficial effect of (fundamental) discords upon consecutive fourths from the bass. The beneficial effect of contrary motion claimed by Mansfield would also correspond to its ranking as an 'exposed' interval next to the fifth. There is no doubt that the fourth is the consonance of third grade and that the fourth from the bass is--to some extent at least—an exposed interval.

On the other hand the rules for its use seem to be far more numerous and stringent than we might have expected from its grading after the fifth. Contrary motion does not seem to be enough to reduce the 'exposure' without the help of conjunct motion in the lower voice or even in both. And if consecutive fourths from the bass are condoned in discords, we should expect to find a single fourth from the bass *à fortiori* tolerable in a discord. But Prout, for example, says (52, 104) that "the rules for approaching and quitting a second inversion apply to the second inversions of discords as well as of concords."

This last statement seems to dispose, not only of the bass fourth as an 'exposed' interval, but as a dissonance as well. Why in approaching and leaving an interval that is merely supposed to be a dissonance should we have to take so much more care than in dealing with intervals that are undoubtedly dissonant? If the rules are too complex to allow us to assume the consonance of the bass fourth, they are just as excessive for its dissonance. The characteristic of dissonances that appears in the fourth is its tendency to suggest the major (or minor) third; it seems to call for a resolution into that interval and to urge the whole chord in which it appears in that direction. But, in contrast to regular dissonance, we find that this characteristic can be suppressed by appropriate circumstances. The discordant feature can be eliminated. In no regular dissonance do we find that any method of approach or departure from the interval will make it appear consonant, however it may alter the trend of its impulse to resolution, or facilitate its appearance in the music.

A trend of resolution appears most strongly in the 'cadential six-four chord,' which is subjected to a rhythmical 'exposure.' The interval is not thereby rendered dissonant; it stands forth clearly as a fourth; we may even suppose that its grading as a consonance emphasises this exposure to some extent; we have no reason to argue that in that case it should be approached by contrary motion, which would reduce the 'exposure'; for its exposure is just what we desire at the moment, seeing that we also give it a rhythmical exposure. But there are only two degrees of the scale that properly invite this cadential resolution—the dominant and the tonic (tonic and subdominant chords). These are the only degrees of the diatonic scale that have a semitone immediately below them. The one gives a cadence on to the dominant, the other on to the tonic. A cadence upon the subdominant is spoilt by the tritone that there appears instead of the fourth. Where tonality is well marked, the dissonantal tendency of the fourth is most useful.

But as it is not desirable to pursue a cadence of this kind whenever a six-four chord is used, means have to be found to avoid the effect. This is done by placing the chord upon an unaccented part of the measure and by leaving it by step in the bass. The other procedure, noted in Table IV, note 1, is not really a variant upon these two. It represents only a temporary movement from the six-four bass.

The bass note of the fourth is, therefore, the most important of the chord, in so far as progression from it is concerned. If it is exposed (1) by being the bass note, (2) by being rhythmically accented, (3) by being the tonic or dominant degree of the scale, it will produce the effect of arrest of motion and strongly suggest the interval of the major third that is so near to it, and so produce the cadence. If such a cadence is not desired, the exposing conditions (3) may, and (2) must, be avoided, and the remaining tendency to revive the cadential tendency must be suppressed by giving the bass a strong melodic force—progression by step.

It is by reference to this tendency towards the third that we must explain the special precautions to be taken in approaching the six-four chord. Even when the bass fourth is skilfully introduced, it still tends more or less strongly to suggest the third. All the more, then, should we expect to find a tendency to confusion of melodic attachments inherent in the approach to a bass fourth. Unless special care is taken the melodic streams would tend to fall towards the third and a jar of surprise would be caused by the fourth actually given. If this jar is to be avoided, we must lead a strong melodic current upon the tones of the bass fourth and especially upon the lower or bass tone which gives the pitch or centre to the whole chord.

Thus we see that the problem of the fourth is akin to that of consecutives. The difficulty in both is the maintenance of clear, unambiguous melodic lines. But they are otherwise different. In consecutives and exposed intervals the obstruction is chiefly caused by the pronounced consonance or dissonance inherent in the single interval itself. In the fourth it is due to the proximity of the fourth to the interval of the third. The distracting influence is external. How it comes to have this effect we shall consider later on.

These difficulties and uncertainties into which we have been led, show us more than ever how desirable it is that information should be gathered about the treatment, not only of consecutive intervals, but also of all single intervals in relation to the circumstances under which they are used, on a statistical basis. To many the statistical method may seem to be a dry as dust business. But if we are to overcome the divergence of opinion which characterises the exposition of harmony and to obtain a body of definite, generally accepted knowledge, it is our only hope. No other method, either, will ever permit us to subtract from the treatment accorded to an interval the part that is probably due to a certain influence so as to leave us with the amount due to any other. These apportionings call for a quantitative treatment, and that is procurable only by statistical methods. At the present time every writer who works out the rules of harmony for himself has to do over again work done by many others before him. And he does not relieve his successors of the necessity of repeating the work. At best he can cover only a small range of the task, and in doing so he is liable to be greatly influenced not only by the generalisations of his predecessors but also by his own special preferences and prejudices. With statistical methods, however, a piece of work, if it is once done thoroughly, is not only finished, but is open to the view of every one else. It need be repeated at most only once for the purpose of verification. Anyone who desires to continue research of a problem already investigated, will have an opportunity in testing the validity of statistics already derived from one composer or period with those to be derived from another. By this means a reliable history of the developments of harmony would in time result. Once the methods of this kind of statistical research were well known, such repetitive tasks might be given to younger students who desire to follow the work of any composer with close analytic attention. As things are at present such a historical view is only present in feeble outline of the broadest kind. We do not know even whether the formulations of the best analysts are complete or how far they still fall short of approximate completeness.

The results we have obtained thus far alone suffice to convince us that with the proper systematic approach and outlook there is every prospect that the science of harmony will one day attain a high grade of precision, if statistical methods are carefully pursued. It will become capable of systematic treatment that should make its apprehension easy and comprehensive. This possibility seems much more probable for music than for the pictorial arts. Music has the advantage of operating with units of sound that are capable of only slight fluctuations from certain forms—their pitches. There are perhaps many people who would look upon such an accurate science of *divinae musicae* as a disastrous calamity. But that is really an absurd point of view. An art can only be furthered by a greater knowledge of its essential nature. Its progress should then be more rapid and sure. We could estimate the possible lines of advance with great probability of success, and if no new vistas seemed likely to open up along our present lines of progress, those who are in search of new lands would know to what point of the system of sounds that leads to music they would have to recede in order to be able to diverge upon strange paths of new outlook. And that would be a great gain. We do not know whether much of the experimentation in music of to-day is not from the outset a waste of time. With a science of music well developed we should be able to judge on this matter with some certainty. The world is not the greater or freer to a genius for his ignorance. He does not create by personal decree, but by discovery of new effects which were already laid down as possibilities in the systematic growth of the art before he took it over. And all unaided discovery is slow and painful, even to a genius. With knowledge discovery may become possible to many others besides the genius, who may then climb nearer to his summits.

In our knowledge of the physical basis of pitch we have a very accurate science of the fundament of music. Here our knowledge is practically complete. But no one supposes that the divine art has become any more earthly for that reason. Why should a science of the art itself degrade it any the more? Those in whom knowledge and its precision tend to dispel the attractions of beautiful and wonderful things[1] will still be able to keep their minds unsullied, if they so desire. But the charm of mystery does not lie in any vagueness of the sensory stuff of art or of its beauty, but in all the longing hopes these finished forms arouse in our minds. We feel the course of life as it might be, were we not our own poor guides stumbling towards ends we can only dimly discern, but the stuff upon which some divine artist had chosen to lay his wondrous hand. We move in the divine thought wrapt up in that stuff of sound and we long to have and to be its life.

Nevertheless the sounds we hear have the precise structure of crystal and their beauty is a chiselled gem. Their sciences may be their equal and counterpart.

[1] Cf. A. E. Hull, *Cyril Scott*, London 1918, p. 78 f.: "Like Debussy, he [Cyril Scott] would protest against the dissection of his music, as if it were a piece of curious clock-work mechanism. In the *Revue Blanche* in 1891 the French master wrote, "As children we were taught to regard the dismemberment of our playthings and toys as a crime of high treason, but these older children still persist in poking their noses where they are not wanted, endeavouring to explain and dissect everything in a cold-blooded way, thus putting an end to all mystery.""

CHAPTER XVII

COMMON CHORDS OR CONCORDANCE

From chapter XI till now we have been engaged essentially in the study of single intervals of two tones. It is true we have considered them generally as they stand in harmony of four or more parts. But our interest centred primarily in the interval of two tones itself, as if it were the element of structure of four part harmony. The results of our study enable us now to show the exact manner in which intervals generally follow one another in harmony of two or more parts.

These results seem both to enrich and to modify the outlook afforded by such previous knowledge as had been systematically sifted. That culminated in the notion of a fusion inherent in each pair of tones themselves and not borrowed from any of their adjuncts. Consonance and dissonance were the opposite poles of this fusion. And consonance seemed obviously to be the ground upon which the pleasures of music mainly stand, although they were evidently greatly enhanced by contrast with dissonance. A general statement of this kind, however, seemed plainly unable to give any sort of adequate expression to the whole nature of music in many parts. The theory showed a crudity and insufficiency very like that of the primitive music of the discantors in comparison with the modern art.

The outlook presented by the concept of fusion was clouded by the emphasis laid upon the approximation of the high grade fusions to the unity and balance of a single pure tone, and the ensuing tendency to carry that notion over into the general idea of consonance and dissonance as they enter into modern music. Thus the function expected from an octave or a fifth in music was such as would express its unity or approximation to the balance of a single tone. The function of a second or a seventh would reveal its tonal duality. But the merest glance at the nature of music seemed to contradict any such conclusion. For in the prototype of musical groupings of tones—in the common chords— we find three essential tones and three intervals. Approximation to the balance of a single tone is out of the question. Even a dull ear would detect plurality in every case. And the interval of the chord most essential to its musical functions is not, as the theory of fusion would most likely suggest, its fifth, but its lower third. But it was not apparent from the previous theory why the lower third should be more important than the upper one. In fact attempts were made to explain the differences of major and minor triads in terms of the reversal of the positions which the thirds occupy in them (43; cf. 60, 35f., 44–53 (Zarlino), 219 ff. (Rameau), 293 ff. (Tartini), 367 ff. (Hauptmann), 385 ff. (Oettingen), 387 ff. (Riemann)). But without avail (cf. 67, 84 ff.; 71, 333; 79). We do not look upon the one chord from below, and upon the other from above.

The great importance of the thirds in modern music did not seem to fit into the theory of fusion. For they were neither high grade consonances nor high grade dissonances. A means of reaching their musical function seemed indeed to ensue upon the distinction of grades of pleasantness in intervals. Thirds and sixths rank high in the scale (29, 194). But it must be evident that mere pleasantness without a justifiable basis of pleasure is a weak reed for any theory of music to lean upon (cf. 71, 351 ff.).

So the outlook upon music seemed to be blocked completely. There seemed to be no means of approach to music as we find it. And it was inevitable that in time an attempt should be made to make a new start, to find a new notion upon which the functions of chords might be grounded.

This idea Stumpf attempted to supply in his notion of concordance as distinguished from consonance (71).

Two notes are consonant when they sound together so as to fuse into an approximate unity, whether the component tones are distinguished or recognised at the same time or not. The greatest degree of unification appears in the octave. It lessens progressively in the fifth, fourth, etc., while in the dissonances we find least of it. Consonance and dissonance appertain in this original and limited sense only to every two tones. "Only as thus understood, as the relation of two tones to one another, is consonance the basal phenomenon of all music" (71, 329). "It must always be borne in mind that what I call fusion can only then be *perceptible as such* when the fusing tones are *distinguished* from one another; just as we cannot perceive similarities without keeping the similars separate. But if this is done, if the three tones of a trichord are distinguished from one another, I at least can form a judgment on their fusion only by pair-wise comparison, but I cannot besides discover a fusion that attaches to the whole, to the triad as such" (71, 330).

It must be evident that thus far at least the empirical teachings of harmony gathered together in the previous pages and the results that have emerged from them confirm this general attitude of Stumpf's towards intervals quite unambiguously. Every pair of tones is in harmony a distinct individual, as it were; it in no way ceases to be itself or changes into another, owing to the simultaneous presence of other tones. As that individual, it carries its own degree of 'fusion' unchangingly about with it, although,—and this must be emphasised,— the effect produced by that fusion at any moment is to some extent modifiable by a number of circumstances other than the fusion itself. Moreover the appreciation of all harmonic effects, even of such elementary ones as we have as yet been able to study, presupposes always in every musical ear some sort of ability to distinguish every pair of voices from every other.

What kind of distinction is implied is left unsaid. Doubtless it may vary greatly in degree of clearness. Low grades of distinction, as musical analysis of finer order would rate them, are apparently quite good enough; for every beginner is supposed to be able to appreciate readily enough what is taught. The ease and certainty of analysis that is habitual in the most finely endowed musical minds is by no means essential. The beginner is not required to be able to name every ordinary chord as soon as heard, or to sing its components, or even to hear them by mental analysis singly, or to separate in turn each pair from the others in his mind's ear. It is enough if he can hear and attend well enough to get the chief effects that are produced by any pair of voices amongst others : e.g. the bad effect of consecutives, exposed intervals, etc. *That*, the teachings of harmony show us implicitly, is already hearing the tones of chords pair by pair[1].

Another point stressed by Stumpf is that "consonance is not changed either by the addition of a third or fourth tone. What is changed is the musical meaning of the tones and their pleasantness. But the unitariness of the octave, the duality of the seventh survives in any and every arrangement with other tones" (71, 328). That, again, is true, but only in the sense of the preceding paragraph. The original essence or being of the consonance or dissonance is not altered, but the effect of it or its suitability or its functions as a unit of musical structure are certainly changed. The terms used by Stumpf—the musical meaning of the tones and their pleasantness—do not specify what these functions

---

[1] Cf. 72, 51-57, which still fails to bridge the gulf between the usual static analysis of tone-masses and the fluid analysis of music.

are. They imply, however, that the degree of fusion is not their basis
or source.

The complicated psychical processes that bring certain modes of apprehension
to bear upon sensations that have been changed either only subliminally or not at
all (often even in a contrary sense) must not be confused with the simple facts of
sense perception by which the basal phenomenon of *all*, even of non-harmonic,
music is given. That one and the same unmodified pair of tones should now fuse
more and now less according as we apprehend it as *c—e♭* or as *c—d♯* is out of the
question, because fusion is a function of the two sensations—or of their physiological
bases—and can change only with these same (71, 328).

In view of the needs and practices of music Stumpf's attitude towards
his notion of fusion is readily intelligible. A generalised notion of fusion
in the sense of degree of unitariness, applicable to any tonal mass of
however many components, would fail to solve the problems of musical
science. Stumpf does well to look about for some new fundamental
notion that will meet the situation. Nevertheless it remains true that
the notion of unitariness is logically quite as applicable to any number
of simultaneous tones as to two. A triad cannot but approximate more
or less to the unitariness of a single tone, even if we add the proviso :
whether its component tones are distinguished or not. If that approximation and that proviso pass for two tones, they must be equally valid
for three or more. Stumpf's attempt to dam up the logical vitality of
the concept of fusion is certainly not the method that will lead us
quickly forwards. The procedure makes a semblance of success only
so long as the waters fail to overflow.

But let us notice the alternative foundations offered :

Our music rests without doubt upon the trichord in its two forms major and
minor. The question then is: what is the objective justification, the reasonable
principle of structure, of trichords? This question is usually either not asked at
all (as in the most of the text-books of harmony) or it is absolved by a reference to
the series of partials. In this series 4 : 5 : 6 are indeed found, and further on the
minor trichord 10 : 12 : 15 as well; but there are in it many other trichords besides,
that are not honoured in such a way by music, although they partly have even
smaller ratios than the minor, such as 7 : 9 : 11. What then gives these two chords
their dominating position, and why must just three tones be bound together generally, if more than one are to be combined at all?

The fundamental principle may be formulated thus : Let the greatest number
of tones within the octave be taken that are severally consonant with one another,
and so that we pass in the tonal motion from below upwards and amongst the
consonances from the stronger to the weaker degrees of consonance.

Starting from any tone we get according to this principle first its upper fifth,
—so, from *c*, *g*, and then only either *e♭* or *e* is further possible, if we neglect for
the present the 'sevens' [5 : 7 and such like]. Thus with the upper finish of the

octave there result the two chords $ce\flat gc^1$ and $cegc^1$. In them all higher grades of fusion are represented. But as it is at once apparent that c^1 has again an octave above itself and within this new octave-space the same process repeats itself, therefore we do not reckon c^1 further as a part of the structure won from c, but as fundamental tone of the analogous one an octave higher. Thus we reach the trichord, in its two forms simultaneously (71, 331 f.).

Stumpf then proceeds to show in a very summary way how the usual chordal combinations of modern music might be developed. Finally he gathers the results together in special concepts.

As a chord we designate a group of simultaneous tones...that can be reduced in the way indicated to chief or accessory triads of a certain fundamental tone. Tone-groups, therefore, with dissonant intervals are called chords in this sense, but not all and sundry, only those that can be obtained from triads by certain operations (71, 337).

Chords, therefore, fall into two classes. Concords (as in our usual sense of the word) must contain a fifth or its inversion a fourth, and a third or a sixth. Discords are all other chords in the sense just expounded. Concordance and discordance are the corresponding abstract terms.

Consonance and dissonance are thus presupposed by the notions of concordance and discordance. But the latter notions differ from the former, which apply only to pairs of tones and to tone groups only in so far as they consist of pairs of tones. Concordance applies primarily only to groups of three or more tones and can be transferred to a tone pair only if and in so far as it is apprehended as a part of a concord, i.e. of a triad (71, 340). Thus one and the same tone pair may be at one moment concordant, at another discordant according to the setting in which it is apprehended. It is the setting that makes the difference. So concordance and discordance only appear with at least three tones. And "consonance is a matter of direct sensory perception, whereas concordance is a matter of apprehension and relational thinking" (71, 341).

Stumpf points out finally that these expressions are not by any means new to musical theory. "But since Franco [the words concordare, discordare have been used] perhaps from the feeling that it is no longer a matter of merely 'sounding together or sounding apart,' but also of 'fitting together and not fitting together'" (71, 350).

Now, however valuable this exposition of the notion of concordance may be in so far as it gives an account of the character peculiar to groups of at least three tones or to intervals as parts of these (compare the notion of 'pattern' expounded in chapter x above), certain points call for immediate remark.

(1) Upon what real ground of tonal functions does the alleged constitutive principle of chords rest? None has been given or even indicated. A logical ground alone is evident. That, of course, is in itself a very important matter, but it is quite powerless to make between tone pairs and triads the real separation that Stumpf claims. It could at most make a merely logical division, such as would divide the discussion or study of tone pairs from that of triads. It could not justify the rule that consonance applies only to tone pairs, concordance only to triads or larger groups of tones. Nor could it do anything to constitute the relational thinking that is claimed as the essence of concordance. It remains as great a mystery as ever how triads with their three tone pairs come to form the basis of modern art.

(2) No doubt Stumpf is firmly convinced that consonance is always a function of two tones at a time. And he may well be right in this. But even then he is so only through 'knowledge by acquaintance,' not through 'knowledge by description.' In other words he feels it or knows it by experience, but he does not know it logically or scientifically. It has not been proved by him. On the contrary the principle upon which concordance is founded would lead us to expect that a chord is only a group of fusional pairs or their derivatives. Then there would be no real division between groups of two, and groups of three or more tones. And there is also no clear reason why we should not turn our relational thought upon a succession of tone pairs as well as upon a sequence of chords or look upon concordant triads as parts of discordances in four or more voices. There seems to be no such radical distinction between music of two parts and music of more than two parts as Stumpf's distinction between consonance and concordance would lead us to suppose.

(3) If two tones necessarily make some approximation to the unity of a single tone of whatever degree, we have still—as far at least as Stumpf's science can show—every reason to expect that any group of tones should do the same. Or rather we should expect that every triad should be rather more dissonant than otherwise. For it would certainly not tempt us to take it for a unity, whether we distinguished its component tones or not.

(4) Here we come upon an important point. Stumpf's dissonance is more or less a negative idea[1], like Helmholtz's consonance in

---

[1] As it is also in many other writers, amongst the ancient Greeks for example, and in Gevaert himself (cf. above, p. 108). But not all the Greek writers neglected the positive aspect of dissonance (v. p. 154, below).

simultaneous intervals. It is merely a minimal degree of approximation to unity, of unitariness (with or without distinction of the component tones or pitches). *Is that enough?* Affirmation would imply that two minimally unifying tones are as such unpleasant. But why so? There is no obvious reason. On the other hand, if while highly unifying pairs give approximation to balance, minimally unifying pairs give, not merely a non-unity, but a positive unbalance or irregular confusion, we should be able to bring that chaos—as a positive ground of unpleasantness—into connexion with similar grounds in other spheres, e.g. pictorial art, logical thought, feeling, etc.

But that is not the end of the subject. Evidence has been brought above to show that there are grades of fusion that must be called neutral—neither distinct consonance nor distinct dissonance. Here we come upon an aspect that does not seem to be subsumable under the fundamental notion of fusion as approximate unitariness. Nor can we well conceive of an indifference-point between balance as approximation to the unity and symmetry of a single tone and unbalance or chaos. How far away from balance should we have to fix this point? But one might say : *consider the middle point to be balance and suppose a departure from it in two directions, one towards loss of balance in unity, the other towards loss of balance in conflict.* At both extremes we tend to lose sight of the component tones. In consonance they run too much into one another, in dissonance they obscure and confuse one another too much.

Such a view would not quite coincide with the useful grading of fusion from a maximum to a minimum on the basis of unitariness. But that would be no insuperable barrier. We might still conserve this grading as a partial aspect of the problem and at the same time prefer the other as more adequate to the sensory stuff. Loss of distinction in unity, balance of distinction, and loss of distinction in confusion can certainly be logically represented as a series from a maximum to a minimum—as a decrease from approximation to the balance of a single tone,—and therefore valid for scientific purposes. But for musical purposes the other notion which centres upon the point of balance of distinction seems by far the more important.

For it simply lays upon our hands the solution of the problem of the great and fundamental importance of the thirds and sixths in all music and of the triad in modern music. The thirds and sixths are the intervals of greatest balance of distinction of tones. Two or more thirds or sixths after one another, therefore, also afford as easy distinction as one. Hence we pass immediately to the interpretation of this distinction as melodic distinction. And so a series of thirds or sixths is most favourable to melodic continuity, as the whole system of facts gathered together in the previous chapters have shown us to be the case.

The importance of the triad for music therefore lies in the two thirds it contains. And, of course, an alternative is created by the two possible positions of the major and minor thirds in each common chord. Thus we get the major chord for the one and the minor chord for the other. Two other possibilities exist, namely the triads containing two minor thirds or two major thirds. These, however, each contain another important interval. The tritone of the one is a distinct dissonance. The (augmented fifth or) minor sixth of the other is not ordinarily a dissonance, but a neutral interval, a maximal balance of distinction. But in the triad it always acts as a dissonance. Many reasons might be suggested for this. We need not attempt to find the most probable one at this point. Having been carried thus far by both fact and logic we may claim to recognise as fact that the fundamental triads both contain a fifth between the outer tones of their two thirds. Upon these two triads all harmony is said to revolve. But the other two triads also are in common use. The distinctive feature of the common chords is due to the fifth they contain. This high grade consonance gives the whole a special unitariness or stability; but this aspect of things we shall leave for special treatment in a later chapter.

The same principle that explains the essence of the common triads will account also for the discords that form so important a part of modern music. If two neutral intervals may be combined to form a triad, it follows as a matter of course that three or more may be combined to form greater chords. These will always be discords. For the repetition at the octave of any one of the components of a common triad gives, as we have seen, merely an extension of the characteristic whole or pattern formed by the three essential tones. Thus we obtain a set of chords in which all the possibilities of combination of thirds may be exhausted. Many other ranges of possible combinations may be taken into view, if all the possible inversions of groups of three thirds are examined[1].

The musical utility of any of these chords will, of course, depend, not

[1] The deduction of these paragraphs is not meant to imply that only chords derived from columns of thirds are to be countenanced. Of this we shall see more as we proceed.

so much upon the neutral nature of the thirds that make up what has been held to be their original position, as upon the kind of intervals that are actually formed between each pair of voices that appear in the chord. It would be a mistake to take any interval or class of intervals as the primary ground of a chord to the disadvantage or depreciation of any other. All the intervals that occur in a chord are of equal importance, except the octave (for the reason we have given). When a seventh or a second occurs, it has the fusional status of a second or of a seventh, and by no means that of the third that may ensue upon its inversion or in relation to some other tone of the chord than the bass of the interval in question.

The view we thus obtain of the part played by thirds in the establishment of chords falls into line with the empirical principle that was extracted by Rameau from the aesthetic work of music and that has been used and defended repeatedly since. This is "the theory of the generation of chords by adding thirds together" (60, 81). It has indeed never been proved in any sense of the term (cf. above, end of chapter x). But it has always made a strong claim to recognition merely by the force naturally inherent in it, apart from all theory, as a generalised expression of empirical practice[1]. And on this ground it must be held to be far more worthy than all the attempts to found a system of chords upon the resonance of the sonorous body or upon the series of partial tones.

Much time and energy has been wasted upon the problem of the fundamental chord or chords from which all the others are derived, upon the systematisation of chords for the purpose of finding their origin, and such like questions. Certainly it was extremely important for musical study to discover the connexions between chords that we know as inversion. That achievement is a piece of direct and unshakable description, which musical theory has to explain in some way or other. And it was equally valuable to work out the differences between systems of intervals and chords, that are summarised in the distinction of major and minor modes. But it is as absurd to put one chord down as the origin of another, as it would be to consider a single interval, or a single tone as the one and only progenitor of all. Besides any chord whatever

---

[1] Well expressed by M. H. Glyn (18, 35): "The third has always been beloved by the natural ear. We have to deal here with a fact of far greater importance to music than any in the science of acoustics, and if consonance to music means the third, and only in a limited degree the fifth and the fourth, while to science it means the fifth and the fourth and after that the third, it is clear that two points of view are being named by the same name which are by nature different and should be recognised as such."

can be reduced to a column of thirds by suitable transposition of its tones through octaves. That follows from the fact that a continuous column of thirds (major *or* minor) soon yields all the tones of the chromatic scale. Thus, $c, e, g, b, d^1, f^1, a^1, c^2 = $ the diatonic scale; the notes of the chromatic scale can be got by suitable substitution of minor for major thirds and conversely.

A musical experimentalist is free to form any interval he can upon his instrument. If it is good and useful he will introduce it wherever it will produce or enhance a desired effect. But he must above all make it possible for the listener to hear it properly; and that enforces the limitation of the number of chords and of their positions in the tonal range. The whole history of music is an attempt to find a system of tones which will yield the greatest variety of chords and the greatest number of relations between them that in turn will most facilitate the apprehension of the tones played and make possible the greatest scope and freedom of aesthetic effects.

It is easier and more natural to bear in mind the actual development of music from the earliest times and to see how the science of musical sounds develops towards more and more complete explanation of its course than it is to try to derive music from an unintelligible genealogy of its final products. The musical mind of the world did not begin under the inspiration of a subconscious appreciation of all musical effects. It began with a very limited sense or feeling for these things. But enjoying what it already had, it strove to make that little grow to greater ends. And as it laboured, the effects were formed experimentally and the ear seized them. The growth of the art gradually revealed more and more subtle aspects and wider and wider connexions or systems of effects. Two of the greatest of these are polyphony and tonality. These were not given; they had to be discovered. Art is as much a process of discovery as science is. Both are experimental and systematic. But while art is content to be empirical, science is restless till it has grasped the whole system of inner bonds that rule its objects and has described them fully and thoroughly.

CHAPTER XVIII

MELODIC MOTION IN RELATION TO DEGREES OF CONSONANCE

THE analysis of the degrees of fusion we have just given may be extended in a way that seems to be of importance.

We have distinguished three chief grades : loss of distinction in unitariness, balance of distinction, and loss of distinction in confusion. And we have noticed that balance of distinction must make for ease and continuity of melody, when two or more melodies run side by side. The question that now promises to further our insight into the structure of music is : what effect has loss of distinction upon melodic continuity, upon the ease and distinctiveness with which two or more melodies will run side by side?

High grade consonance we have already learnt to look upon as balance, approximation to the unity of a single tone. Its component tones are wrought together more than usual; they cling together and do not offer to pass with as much ease into two other tones as they were approached from those that preceded them. They constitute, therefore, a point of relative rest and tend to bring the voices to a stop.

The same effect is produced under certain circumstances by the unison, as we have seen above (p. 130). A unison is, of course, from an absolute acoustical point of view a single tone, certainly not an interval. And a single tone is not as such in music arrestive in function. Unison has a significant meaning only in so far as two melodic streams are felt to meet and to be identical in a certain sound. Being usually two, they then 'sound one.' When such a unison is presented without correction by the various circumstances brought to bear upon high grade consonances to make them mobile, it produces the same undesirable effect as they do. Two voices are caught up into one and an effect of arrest and confusion is produced by the loss of distinction in unity. It is only in this way that the functional similarity of unison and high grade consonances can be justified.

This function of stability and arrest[1] peculiar to the consonances of the octave and fifth has long been recognised and shows itself in various ways. In Greek and early Western music the octave (or unison) was the usual close of a piece. The root position of a triad is a more

---

[1] Or 'repose': cf. D. F. Tovey, 74 *passim*.

stable form than either of its inversions; the fifth in it spans the two thirds or gives at least (in the alternative arrangement) a fifth with the bass. In the first inversion there is either no fifth at all or in the alternative arrangement only between the two upper voices. (Conjunction with the bass we have already seen to be generally more powerful than conjunction with the soprano.) In the second inversion we find the peculiar feature of a fourth from the bass. As a consonant interval that will produce some effect of rest corresponding to its grade, which approximates towards the neutral range, though still above it. Of these three forms of the common chord the first is, as the older theorists said, 'most apt to conclude'; it produces the greatest arrestive effect upon the flow of voices. When this effect is heightened by special means—by the use of the chief transitions of tonality, dominant or subdominant to tonic, and by suitable rhythmical exposure, etc., we get the various cadences, whose sole function is to produce partial or complete arrest. Other contributory, and therefore functionally similar, conditions are a gradual reduction of speed, a greater sonorousness and steadiness in tone production, the repetition of the final chord, and so on.

The counterpart of the high grade consonances is formed by the high grade dissonances. These create a loss of distinction in confusion, which must also have a disturbing effect upon the ease and continuity of melody. The treatment of dissonances in music is a natural consequence of this. If dissonances are to be introduced, special means must be employed to overcome the loss of distinction as far as possible. Devices, such as suspension and preparation, were early invented and rigorously prescribed. Although they are not now considered to be indispensable, they have by no means been superseded as superfluous. They serve to fix in advance in the hearer's mind the most difficult part of the group of tones about to be presented, whereupon the others are introduced by common and easy melodic procedure. In this way the listener is enabled to follow the movement of all the melodies equally well, and so is shielded from the confusion that might otherwise arise. Besides this subtractive method of approaching dissonances, there is of course the method of contrary motion by which the tones forming a dissonance with one another may be approached from opposite sides. The listener is thus guided carefully through the moment of confusion.

The musical mind of to-day has grown so familiar with all the melodic combinations our harmonic procedure has reduced to distinct types, that there are many who almost suggest that in the course of time, as music progresses, what was previously discord comes to be reckoned as concord. Stumpf said : "Effect upon feeling is specially liable to change, even within our system, in that the unpleasantness of discords weakens and through the introduction of new and ever bolder discordant structures the old ones take on the feeling effects of concords; so that, as v. Hornbostel remarked, progressions to these old discords can act soothingly like a resolution into concords" (71, 341 f.). Stumpf, however, does not think that such changes of feeling will ever break down the difference between concord and discord. We may well agree with him; for a definite reason can be given that seems to be of substantial weight.

However familiar we may become with the patterns of discords, that will surely never in any way alter the graded differences there are between consonances, neutral fusions, and dissonances in respect of balance of tonal distinction. Both the current theory of the derivation of all intervals from the series of partials and Stumpf's theory of fusion grade the intervals in a series of indefinitely decreasing consonance. The one end of Stumpf's series, as we have seen, is characterised by apparent unitariness of sound, the other by closer and closer approximation to mere clear-cut apprehension of two-ness. The theory of partials suggests that the nearest and loudest and perhaps most frequent partials yield the consonances that are distinguished earliest in the history of music, and that as music advances our familiarity with partials extends farther along the series, so that we reckon as consonances always as much as we have thus made our own. The history of music seems to provide a parallel to this in the early use of the octave, the subsequent 'organising' in fifths and fourths, and the later classification of thirds and sixths, or even sevenths as consonances. But, on our interpretation, all this line of speculation is completely cut out. The series of fusions has its neutral point—or its region of natural ease and familiarity, as it were—in the middle, in the thirds and sixths. From this point the difficulty of manipulating the intervals in polyphony or in harmonic music increases in the two opposite directions—towards the consonances and towards the dissonances. Familiarity may give us greater facility in dealing with these naturally recalcitrant intervals; it may even induce us to dispense with certain aids to apprehension that we once found necessary or desirable. But it cannot alter the natural differences between the various grades.

Once we have found the true system of functions of intervals, the false motive suggested by the apparent course of history entirely loses its value. The historical order of approach is quite irrelevant and can be readily explained otherwise. It was the prevalence of monophony that led to the adoption of the octave first of all intervals; monophony does not essentially change in becoming homophony. And homophonies in fifths and fourths are the next inevitable attempts at continuous development. The great consonances call early attention to themselves. But it is only in polyphony that the polyphonic functions of these and all the other intervals can be discovered. And it is these functions of intervals—whereby they either yield a simultaneity of easy flowing melodies or disturb one another in this respect—that determine the final classification of intervals.

As for the plea that the Greeks had not yet recognised the consonance of the thirds and sixths, the evidence seems to indicate merely that they did not reckon these intervals among the distinct consonances. Neither do we really. They do not show a notable degree of approximation to unity ("so that the resulting sound is one like and similar to a single one," Nicomachus—71, 329; 66, 54). That still leaves room for two other classes, one in which the two sounds, far from being one-like, are rather specially two-like,—shall we say?—or *dia*phonic[1], *dis*cordant, contraposed ("when the sound of the two is as it were rent asunder and without true blending," Nicomachus—66, 54); and another middle one in which the two sounds are just two, neither friends nor enemies, but just comrades.

Knowledge by acquaintance may change then; and so may knowledge by theory, and practice, and familiarity and all such adjuncts of feeling or sensory experience; but sensory feeling itself does not seem to change. The conformations of sense retain their characters unaltered. Sense is a stuff that the growing mind of man may learn to mould as he can, but ever in obedience to the laws inherent in it. It is as much an objective world that we must learn to know and to use as is the world of nature.

---

[1] Cf. 14, 96: "La sensation auditive produite par les consonnances et les dissonances est analysée d'une manière uniforme par tous les écrivains: "dans la consonnance les deux sons se mélangent au point de s'absorber mutuellement, de telle manière que l'oreille ne reçoive qu'une impression unique, douce et suave." Elien le platonicien compare la consonnance à "du vin melé de miel, ou aucune des deux substances ne prédomine, et ayant le goût d'un breuvage particulier, qui n'est ni du miel ni du vin. Dans la dissonance, au contraire, le mélange ne s'opère pas; les sons se repoussent, pour ainsi dire, l'un l'autre, et l'impression totale est dure et pénible." Of course the idea in "ni du miel ni du vin" must not be pushed to the extreme of positing a new resultant third tone (cf. 16, 138 f.; 66, 52). "Aussi les définitions antiques de la symphonie et de la diaphonie sont-elles en grande partie sanctionées par l'oreille moderne" (16, 138).

But to pass on. Dissonance must differ from consonance not only in the way described, but in another subtle manner. If consonance creates the effect of a pause or rest by presenting us with an approximation to unity, and if we accept the suggestion to rest, or if it does not conflict with the effect of the other tendencies of sound at the moment, but forms a consistent whole with them, then we shall be somewhat careless of distinguishing differences within the unitariness. We are then either wholly or relatively at rest and we do not need to be scrupulous in distinguishing. We have no need for such finished distinctions, for we do not crave to move forwards. If, on the other hand, we want the music to move forwards through a consonance, in spite of the tendency of consonance to create repose, we must be careful not to strengthen the reposeful effect by any means of the same tendency. Hence flow the rules against consecutive octaves and fifths, and the exposure of these intervals.

In dissonances, on the contrary, a point of unrest is created. There is neither rest nor even flow, but a disruptive effect, and disagreement between the component tones. They get in each other's way and produce mutual confusion and disturbance. Even if we have been led skilfully into the dissonance, we nevertheless are impelled forwards. We look for another phase of progress in which disturbance shall cease. But not any consonance (or dissonance) may follow; only one which we can easily and melodically reach from the present dissonance. Thus the need for resolution of dissonances arises. When several voices run concurrently, and there is consequently more to follow and more danger of losing the thread of sequence, the need for resolution is all the greater, and must be the more carefully controlled. Thus it appears that dissonance, far from being a barrier and a hindrance to good music, acts clearly as a stimulant upon melodic activity, urging it forwards and increasing expectation of progress.

Much might be said in favour of the adoption of the common Greek terms 'symphony' and 'diaphony' with the addition of the rarer term 'paraphony.' In symphony the tones of an interval tend to become indistinguishable through too much unitariness or fusion; in diaphony they sound through or against one another, disturbing and confusing one another; in paraphony there is balance, so that melodies formed of such intervals will flow evenly side by side, the one not inhibiting the apprehension of the other. Paraphony, it should be noted, does not imply that the tones of such an interval are on the whole more easily apprehended as mere duality than the tones of a diaphony. In the latter the duality is indirectly emphasised by the harsh confusion. Paraphony implies merely a medium grade of obviousness of plurality through lack of approximation to the unity of a single tone (in the sense of Stumpf's fusion), but at the same time a maximum grade of distinguishability for those musical purposes which require distinction of pitches and apprehension of melodic flow in more voices than one. For these purposes we require a perfectly clear-cut untroubled distinguishability of the component tones of an interval. That is plainly wanting at the consonantal end of the fusional series. It is merely presumed to be present at the dissonantal end, because at that end there is the more obvious indication of two-ness of tone; the tones jar upon one another harshly and there may be very obvious beating between them; even the unmusical mind reads there signs confidently as two-ness, while the musical ear finds in the dissonances more obvious lack of fusion or presence of ruption than in the thirds and sixths. But it would be right to claim that in the small dissonances it is more difficult to pick out the pitches of the component tones than it is in the thirds and sixths, and that in the larger dissonances, although the pitches stand well apart, yet the two tones do 'oppose' one another or jangle with one another[1], besides being very much two-like apart from that. Thus the validity of the fusional series, as we find it in Stumpf and others, would be called in question. In music an abstraction of melodic function from apparent one-ness and jarring two-ness has been carried through, which points to the neutral thirds and sixths as the region of pure two-ness of tone in interval. The clear distinction of melodies must therefore be but a seriation of this two-ness function, not a new kind of (perpendicular) distinction of tones supervening upon their melodic combination. It is a task for experimental work to find and to describe exactly the analytic attitude that will confirm this conclusion from the functions of intervals in polyphonic music.

Our new results may be summarised very briefly with the aid of these terms. The movement in music is melodic. For the proper flow of simultaneous melodies intervals must either be themselves actually paraphonic or they must be used paraphonically.

Note.—Of the Greek terms relating to harmony, symphony and diaphony are by far the most familiar and have gone over simply

---

[1] In particular the lower boundary of the higher tone falls near to, and so defiles, the symmetrical outline of the lower tone in its most important point—its pitch.

into the terms consonance and dissonance. I have quoted characteristic examples of their definitions above (p. 154). The term paraphony was used by several later writers, Thrasyllus, Bacchius and Gaudentius (66, 48 f., 67 ff.; 16, 139). While the relevant passage in Bacchius is almost certainly confused, Thrasyllus attached the term to the fifth and fourth in distinction from the octave, a subdivision which does not seem to be of any particular interest, as far as we are concerned nowadays. But the relative passage in Gaudentius is of the greatest importance. It has apparently been a source of mystification for most interpreters. Stumpf speaks of it as "this otherwise (i.e. than by his explanation) quite incomprehensible passage" (66, 72)[1].

Since my conclusions call so clearly for the use of a term like 'paraphony,' it becomes a matter of much interest to consider whether Gaudentius may have been led to his somewhat similar application of the same term from a similar train of thought. Apart from the mere existence of the term, its verbal meaning, and the conceptual setting of the notion between symphony and diaphony, I can find with the help of the chief authorities (Gevaert and Stumpf) little or nothing to show what the idea of Gaudentius really was. Stumpf suggests merely that Gaudentius meant the major third and the tritone to be taken simply as "consonances of lower grade, as transition to the dissonances" (66, 70).

Stumpf's translation of the passage in Gaudentius is as follows :

Symphonic are those in which, when they are simultaneously struck or blown on the flute, the melos of the lower in relation to the higher or conversely is always the same, or (in which) as it were a fusion in the performance of two tones occurs and a kind of unity results. Diaphonic are those in which when they are simultaneously struck or blown, nothing of the melos of the lower in relation to the higher or conversely appears to be the same or which show no sort of fusion in relation to one another. Paraphonic are those that, standing in the middle between the symphonic and the diaphonic, yet appear symphonic when played; which seems to be the case in the tritone (f–b) and the ditone (g–b), (66, 69).

Elsewhere (66, 66) Stumpf says that melos is "perhaps best translated as 'the melodic element of tone' or the tonal element of melody."

No doubt the matter might be discussed at great length. But if we leave the first clauses of Gaudentius's definitions as obscure, and (or) interpret them by the following clauses, then we must look upon Gaudentius's use of the term 'paraphony' as well founded. Symphony is the greatest fusion (when the melos of the two tones tends to sameness),

---

[1] Cf. 14, 99: "Mais il importe de remarquer que cette doctrine est isolée dans la littérature musicale des anciens."

diaphony is the least fusion (when the melos of the two tends to extreme difference), paraphony is the middle between these, which appears as symphony in performance. It is probably this last clause that has compelled writers to think Gaudentius meant this middle class to be a lower grade of consonance, not a really middle neutral class, and so to miss perhaps the main point of Gaudentius's distinction. That main point is the placing of the neutral relation (para) in the middle of the whole series and the attaching of it in particular to the major third. It is much more significant than any distinction between the octave and the fifth with the fourth. Music naturally gives the octave a special place, because it is its unit of division, whereby repetitions in octaves become practically mere repetitions or identities. But the grading of octave, fifth, fourth had been established long before Gaudentius (by Aristoxenus, 66, 38). The former did not add the major third to them as a lower grade of consonance, but as a member of another class lying between the consonances and the dissonances.

It may seem perverse to stress the point; but it is really an important one. Music makes a distinction of opposition between consonance and dissonance, but it has failed to recognise that distinction theoretically in so far as it ranks the thirds and sixths as *imperfect* consonances, not as neither consonances nor dissonances but neutral 'sonances' or paraphonies. Gaudentius makes this division, adding the note that paraphonies sound symphonic in (instrumental) performance (ἐν τῇ κρούσει).

In a sense, our present musical classification may be said to be just the obverse of the early Greek one, primitive and limited as that has been usually thought to be by modern writers. The older Greeks made one dividing line below the fourth, calling all the rest diaphonies. We draw our great line below the thirds and sixths, calling all except the dissonances consonances. Let us combine the two and we get—in principle—the divisions of Gaudentius. And if we like, we can add with him that ἐν τῇ κρούσει—in a mere interval of two tones struck together as distinct from two melodies (as we may say)—the paraphonies seem symphonic. In other words when it is not a matter of paraphony generally, but just of the general character of a chord as a whole, we reckon the specific paraphonies (thirds and sixths) to the symphonies. Or, we might say : we take them not as apparently one-like like the high grade consonances, but as agreeing with one another and therefore as consonant, because they do not obviously jar upon one another as do the distinct dissonances.

Of course it would be improper to read into Gaudentius all that we can now put into the skeleton of the distinctions to fill them out. But he should have the benefit of any doubt there may be. His conceptual scheme is full enough, but there is not enough detail.

All modern theorists have treated Gaudentius's distinction as if it merely amounted to an extension of the grades within the class of consonances by a further step downwards. That surely does violence to his words and to his term (paraphony). If this modern view is wrong, then with it must go the attempt to see an evolution of the notion of consonance downwards from the octave, to include first the fifth and fourth, then the thirds and sixths, now the natural seventh and tritone, and to-morrow all the dissonances themselves (cf. 22, 115). Impossible! That were no evolution, but a debasement. Evolution—unless it be the degeneration of the parasite that casts off its sense-organs—means progress, an increase in the complexities or in the differences distinguished, not the swamping of all differences in one class[1]. All differences remain as they were given, but we learn to know them and their functions better, and to use them practically in our art without feeling shocked or lost amongst the more refractory ones[2].

[1] So the distinctions made by such a writer as Johannes de Garlandia show rather a keen sense, than a 'parade' (81, 158) of scientific accuracy. His series is: unison, octave; 5, 4; 3, III; VI, 7; II, 6; 2, T, VII. Cf. 6, vol. I. 104 f.

[2] It may be of some objective interest to note that I made a reference to Gaudentius on pages 15, 108 above, but at that time I saw (like all other students of the subject, I suppose) nothing specially significant in the term paraphony. In fact I had in the meantime till writing the present section forgotten the existence of this term. The train of thought expounded above made me feel the need for a term to cover the range between consonance and dissonance, but different from either. Latin is here insufficient unless we say sonance; so I put together the term paraphony. Months afterwards on turning over the pages of Gevaert I was astonished and delighted to see the word paraphony there. The point of interest is that whether unconscious cerebration was agog in this coincidence—which I greatly doubt—or not, the objective differences analysed and discussed demanded the term; it was not suggested by, or transferred from, Gaudentius. So perhaps my analysis will do him a good turn for the dignity he gives to the term I chanced upon.

CHAPTER XIX

MELODY (OR PARAPHONY) AS THE PRIMARY BASIS OF MUSIC

WE have now reached a point of view from which we can survey a large part of the realm of music.

The view we obtain has been clearly indicated in the thesis which forms the title of this chapter. Melody is the primary basis of all music. By melody we mean a special phenomenon of motion or passage between two tones that appear before a mind in successive moments separated from one another by a certain interval of time which may vary in size within certain limits under various conditions. The successive tones must not be so different (in loudness and blend) from one another as to appear to come from different sources and so to suggest an objective independence of one another. That circumstance is unfavourable to melodic connexion. But it is not our present concern to study the nature of the motion involved in melody, or in short of melody in general. We have here taken melody for granted, as a familiar phenomenal fact. The reader is supposed to know already what melodic connexion is, as he surely does, being able to tell at once whether the 'passage' from one note to another is there or whether there is a break, a suspension of motion or passage, as there is for example after a close or a half close, etc. (For an account of the primary theoretical study of melody in this sense, see 77, chapter VI.)

Melody, in short, is the motion of music.

But the word is often used to mean more than that, namely the series of pitches through which a melodic motion passes. For general theoretical purposes it is best to use for this a word that links this feature of sound to the analogous feature in the other senses. When a visual motion passes through a number of points, we say it marks out a certain form or figure. Thus a burning torch swung quickly round leaves a trail that forms more or less of the circumference of a circle. A flying meteor marks out a straight line, and so on. We may say similarly that a melody falls into, or has, a certain form or figure. The notion is quite familiar to musical literature. Perhaps the word 'theme' and its derivatives 'thematic' and 'thematised' are less open to confusion with heterogeneous subjects than are any of the other words that bear a similar meaning, such as tune, motive, subject, etc.

If we use the term 'melody' for the general notion of motion from note to note, then we may divide melodies into two classes—those that are thematic or that show a definite form capable of coherent analysis, and those that are not thematic. In the latter there may be plenty of motion from tone to tone, i.e. plenty of melody in general, but as little form as there is in the motion of a fly—to take a homely instance. The fly is always on the move, and so is a dancer. But there is figure or form in the dancer's actions, while in the fly's there is practically none.

In this sense melody is the primary basis of all music.

With so general a meaning in the term melody, this statement partakes very much of the nature of a truism. Music undoubtedly began as melody. Apart from purely rhythmic art in which sound plays only the part of a practically unvaried medium, all early music is simply melodic. And it is thematic in a more or less simple way as well. But it is generally supposed to have ceased to be wholly or thoroughly melodic at a certain point of its development in Europe, and to have become harmonic; and in so changing to have struck into a new line of development that was present only in minute traces, if at all, in primitive or in ancient music. This new line has led to wonderful forms of art, unshadowed and undreamt in the first origins of music. Harmony seems to be a new creation within music; a new dimension, one might say; the perpendicular complement to the horizontal functions of melody, as has been said.

No doubt harmony has come to be of great importance. It is not easy, however, to say precisely what its scope is. But there seems little doubt that the melodic functions of music have been considerably underestimated. In fact, from the point of view of theory, harmony has usually been put down as the one and only basis of true music. A music in which harmony is evidently neither implicit as in polyphony, nor explicit as in harmonic music, hardly deserves the name of art. It is merely primitive play, as it were.

An almost contrary thesis may be vigorously maintained. It may be claimed that melody is the primary and continuous basis of music; remaining so even throughout harmonic developments, which are essentially a by-product of melodic complexity, always carefully subordinated to the prior and essential requirements of melodic movement.

In the earliest music there is only one line of melody, only one voice. Harmony shows itself at most only in the reduplication of this melody at the octave, fifth, or fourth, or in an irregular accompaniment of intervals to the tones of the melody which does not make a second melody and does not seem to interfere with the apprehension of the one distinct melody. From this somewhat chaotic state the art of polyphony in two and then in three or more voices gradually emerges in that the accompanying tones take on the form of a distinct voice chiefly by the device of contrary motion with the chief voice[1]. Then the art of polyphony becomes clearly conscious and slowly attains a sense of the true principles of style suitable to simultaneous melodies (cf. 48, chapter II). The rules regarding consecutive and exposed intervals are then gradually discovered. These rules are necessary in order that the melodic distinction of the voices may be clear and easy. The arrestive effect of the 'symphonies' is of the greatest importance where arrest of melodic progression is desired; but it must be carefully avoided when melodies have to flow easily together. Diaphonies must not lead to the confusion of melodies; we must be led safely through them and on to groupings of tones that the ear can readily apprehend.

But there is in this kind of art no difference, except that of difficulty, between the construction of two simultaneous melodies and of three or more. Harmony does not necessarily become explicit with the grouping of three voices; nor does music in two voices (now) necessarily fail to be apprehended harmonically; there are many who claim that even a single melody is necessarily apprehended harmonically, absurd though that claim—in the face of primitive music—must be held to be. Of course polyphony can hardly begin to be apart from a cogently melodic or even thematic treatment of the combined melodies. That follows naturally from their very being as distinct melodies. There would be as little interest or beauty in the erratic or merely melodic motion of several voices as there would be for us in a song made up of a more or less random succession of tones[2].

[1] It was Helmholtz that suggested that "the first of such examples could scarcely have been intended for more than musical tricks to amuse social meetings. It was a new and amusing discovery that two totally independent melodies might be sung together and yet sound well" (20, 244). But the actual course of development must have been more natural and continuously meaningful than that.

[2] C. V. Stanford (62, 6) extends this idea in a general rule with considerable emphasis: "Mere combinations of notes, in themselves sounding well, but without logical connexion with their successors, are useless as music. The simultaneous presentation of two melodies which fit each other is at once a musical invention; and when a third or fourth melody is added to the combination, the result is what is called harmony. To speak of studying harmony and counterpoint is, therefore, to put the cart before the horse. It is counterpoint which develops harmony and there is no such boundary wall between the two studies as most students imagine." Our conclusions re-affirm this last statement.

Out of polyphony harmony was gradually abstracted by a slow process of association of simultaneous tones and familiarisation with their perpendicular aspects. The lines of melodic motion had to meet often and often ere the patterns of their contacts could become thoroughly known. And still longer time was required before men found that these patterns could be ranged in succession in most fascinating ways. This origin of harmony is in a sense quite familiar. It has been well expressed by Sir Hubert Parry in his articles in Grove's *Dictionary of Music and Musicians*. At the end of the article on Harmony he speaks of it summarily as "the principle that harmony is the result of combined melodies."

The ecclesiastical cadences were nominally defined by the progressions of the individual voices, and the fact of their collectively giving the ordinary Dominant Cadences in a large proportion of instances was not the result of principle, but in point of fact an accident. The Dominant Harmonic Cadence is the passage of the mass of the harmony of the Dominant into the mass of the Tonic" (46, 310).

But harmonic music has not ceased to be essentially melodic. The thematisation characteristic of polyphony has disappeared from the majority of the voices perhaps; but each chord must still be connected with the preceding in ways which make melodic connexion easy and which do not allow its clear flow to be arrested by excessive symphony or dissipated by extreme diaphony. The paraphonic components of chords are perhaps their essential musical constituents. And the discovery of new chords may be said to be the development of new paraphonic combinations. Most of the modern systems of chord formation have emphasised the importance of the superposition of thirds as a principle of origin. We can now understand it not merely as an empirical principle, but as a principle of function (in so far, of course, as it rests not upon a mere arithmetic of thirds by transposition through octaves, but upon a naturally felt, and so real, connexion through the similarities of inversions to one another as patterns). The primary need is not so much a wealth of consonantal and dissonantal effects, as a clear and intelligible flow of melody through such harmonic effects as are compatible therewith.

"By the use of chromatic passing and preliminary notes," Sir Hubert Parry says, "by retardations, and by simple chromatic alterations of the notes of chords according to their melodic significance, combinations are arrived at such as puzzled and do continue to puzzle theorists who regard harmony as so many unchangeable lumps of chords which cannot be admitted in music unless a fundamental bass can be found for them" (46, 318; cf. 47, and 48, 234).

This interpretation of the development of chords is to be preferred to Stumpf's. It is certainly not necessary to introduce the notion of concordance in order merely to explain the conjunction of more voices than two. The notions of consonance, or better, of paraphony—as may perhaps be said generically—are quite adequate to the use of intervals, not only between two voices, but also between more than two. No change of basis or of principle is thereby required. The essential basis of music ever was and remains melodic movement. One melody is self-contained; simultaneous melodies must be mutually compatible. Single melodies are almost inevitably thematic; simultaneous melodies may be so too; but the thematisation of one or more may be dropped in favour of the interest created by their harmonic fusion. Or, finally, as D. F. Tovey expresses it, modern melody may be merely (is) "the surface of a series of harmonies" (75)[1]. When harmony emerges, however, it is not a new creation; it is still precisely the same thing as is the fusion or balance of two tones, except, of course, that it is of greater scope and detail in several voices than in two. It has now only been made the centre and object of the artist's creative genius. The art has become perpendicular in build instead of horizontal.

But the rejection of harmony or concordance as an essentially new element in polyphony does not imply that the latter in its harmonic form has brought forth nothing. There has certainly been great development and growth, so much indeed as to create striking differences in the styles of art. The interest created by these was so great that much of what had been toilfully gained in polyphonic art was temporarily abandoned and fell into common neglect. That must perhaps always happen to any art when new constructive vistas appear.

There seems to be no doubt that of the acquisitions of the new art the most important and fundamental was the principle of tonality. In polyphony the thematisation of all the melodies held them artistically together. This common structure could only be abandoned when a new principle of connexion had become apparent, something that would link the various voices together throughout the changes of their harmonic patterns. These patterns had to be wrought into an intelligible system, and, as we know, even the outline of the scales had to be altered to make this possible.

---

[1] Or, "modern melody is the musical surface of rhythm, harmony, form, and instrumentation. In short melody is the surface of music." This may be the modern culmination of melody, but it is certainly a wrong definition of melody in general, including primitive melody.

It is tonality that gave and continues to give the chief impulse to the systematisation of chords. These are, and perhaps can only be, systematised in relation to tonality, and its three poles—tonic, dominant and subdominant. It is a perversion of actual history to suppose that chords were first derived from the mere tone (fundamental and partials), bringing implicit in them the determinants of tonality and of our scales. Scales were developed long before either tonality or chords had been conceived or even felt. On the contrary the course portrayed in history seems systematically much more acceptable. First came mere motion in tones (without any implicit scale) or mere melody; out of melody was begotten by the force of familiarity and the needs of social co-operation scale; scale made further complications and co-operations possible, leading to polyphony; the habits and traditions of polyphony engendered harmony and tonality; and that as it grew reflected upon its progenitors and moulded them to its better development; it adjusted the scale to its systematic requirements and reduced the functions of thematised melody in favour of shorter bonds of melody running through and combining large tonal masses. Thus line was reduced in favour of mass; but the mass itself is now also treated linearly, as it were; or it is made to move round a central axis of orientation. It is only after this time that we find the first beginning of a systematic exposition of chords in Rameau. And we must remember that, curiously enough, one of the first steps in this work was the identification of those groups of tones that we know as inversions of one another.

We have already discussed the problem of inversions and have introduced the notion of 'volumic pattern' to account for their connexions and differences. Possibly the chordal aspect of modern music is founded upon this attitude towards volume, whereby the group of tones is considered rather as a whole than as so many stages in so many melodies of which the themes are clearly held in mind. Of course the mind has still to be led melodically through these masses. But the interest lies not in the forms created by the melodic movements as such, but in the masses, their patterns or 'surfaces' or 'colours' (as some say) into which the mind has been easily and safely led. The modern interest would lie, then, not so much in the forms of all the motions through which the listener is carried, as in the phases or masses of sound in which he from moment to moment finds himself.

The canvas is now filled with broader effects. These are still all built upon lines of movement (i.e. melodies), but the lines are now subdued and hardly appear from a distance, so to speak, except it may be in the one, or perhaps two, that run through the mass in its changes and give its movements thematic form and beauty.

Tonality may be said to be perhaps the broadest of these effects. In its earliest forms it is merely a centre or point for the harmonic stream. In its later and freer developments it might almost be considered as a new form of movement—a new melody—of all the (unthematised) voices at once and it builds itself up like the primitive scale upon the simplest consonances, of which the fifth is, of course, the first distinguishable one,—the octave being by its function already mere identity or repetition. So we get the poles of tonality,—the dominant and subdominant. And these again give others, equally closely allied to them, until a whole system appears through which the harmonic stream may be made to wander for its greater diversity and beauty.

One of the perennial problems of tonality has been the nature and origin of the difference between the major and the minor scales. Their actual constituents, as we have noted, have been determined by the systematic requirements of the groupings of the moving voices that are most essential to them. In their most rudimentary form these reduce entirely to the two thirds (and sixths)—the two chief inherently paraphonic intervals. The tonal difference between the major and minor keys may then be read from their symbols of 'origin' in the two common chords ceg and $ce\flat g$.

The different relations of these two to the partials of their supposed root c have been grossly overrated in importance. Certainly, if c has a full series of lower partials, e and g will coincide with them. But, as has so often been pointed out, the partials of e and g,—which cannot be ignored,—confuse the issue. For e gives $g\sharp$ and b, while g gives b and d. As Macfarren said :

It is of course necessary for practical musical purposes, not only to make a selection of notes from the endless harmonic series, but to confine the use of harmonics to those belonging to certain exceptional generators, or roots, in every key; otherwise every note of every chord might be supposed to furnish its harmonic series, and each of these its harmonics in turn, all sounds would confuse all other sounds, tonality would be at an end, and Babel would reign supreme (35, 94).

In spite of the fancies of recent composers in search of new scales (23, 72f., and *passim*; cf. 5, 24), these harmonic origins have for many a long year obviously been as dead as a door-nail. A much simpler origin of such latest varieties is their actual source—the chromatic scale of just or of equal temperament, consciously treated by the process of 'aesthetic selection' that Helmholtz emphasised. No doubt, however, the coincidence of partials will make for slight differences of smoothness; these we have already admitted as modifications of smoothness otherwise given.

Paraphonically the two common chords show practically no difference at all. Both contain a major and a minor third and a fifth. The major third is 'exposed' in the major chord by its resting on the bass voice; the minor third is similarly 'exposed' in the other. Thus the two chords are practically equivalent.

The only difference that remains is the difference of pitch of the middle tone. It lies lower in the minor chord than in the major. And this difference must run equally throughout the whole of the two scales. Any composition distorted from its major tonality into the tonic minor differs from the former solely in the lowering of all the mediants and submediants by a semitone. The minor form differs from the major by an inner lowering of pitch. It is more 'voluminous,' heavier, darker, sadder, etc. Change to the major key means a streak of lesser volumes, a brighter, lighter, clearer atmosphere, as it were[1].

When the problem is thus cleared of its false and artificial difficulties, the solution is easy and it appears to be quite natural and inevitable. There is no special agony of harmonic birth in the minor; we do not look upon the major from below and upon the minor from above; they are not the mirror images of one another. The minor tonality partakes merely of that difference that appears in the raising of the pitch niveau of a composition. Smaller volumes suggest brighter, lighter effects; or they merely *are* smaller, more precise, tones, apart from all suggestion. And so the minor key appears woven with one larger thread throughout; our minds respond to this auditory difference with any analogously varied experiences we may have ready in our memories and moods[2].

It is a familiar fact that in recent years many experiments have

[1] Cf. the detailed empirical investigation of Becker's, where it is shown that "Dur bedeutet schlechthin Lust, moll schlechthin Unlust" (2, 255); and (p. 258): "As expressions of pleasure in the minor are not free of unpleasant moments, while on the contrary major as expression of displeasure approximates by a greater or less indefiniteness of coloration to the minor character, and, besides, the displeasure value of many contents is not very distinct, therefore we can uphold the meaning of the modes as opposite feeling-tones of corresponding musical devices."

[2] Contrariwise and similarly, for the ancient Greeks "the passage of the minor third into the major expressed a lowering, a depression" (16, 324). For in their music the melody lay *below* the accompaniment, so that the major third (reckoned downwards), gave the touch of weight or sadness. On thirds in Greek music see also 16, 266.

been made by composers towards the establishment of new schemes
of melody and harmony. It is unnecessary to specify or to describe
these here. Time must first show what is their capacity for lasting
artistic usage. But it is evident that the basis of music expounded in
the preceding chapters is quite ready and able to accommodate all that
is proved to be acceptable in them. If our theoretical analyses and
constructions correspond fully to the chief long tried and approved
forms of music, they·will apply to the newer departures in their degree
of success. For these actually hold already for the musical mind a
very definite relation to the earlier established music. We hear and
know what we gain in the new and what we lose in it that made the
old precious. There is in all a gradual transition and development.
The mind may grow familiar with the natural and artistically achieved
paraphonies of the main drift of musical art and tire of them. It may
then strive to curb the lesser paraphonies to its will. But the gain of
novelty in chords of fourths must involve a great strain upon the
apprehension. Many groupings of lesser paraphonies must be in actual
effect much more diaphonic than paraphonic, discordant more than
melodious. But it is not our present intention to defend or to justify
any one of such experiments, but merely to show that the basis of
analysis and theory already offered not only leaves room for them,
but can even anticipate the losses and deficiencies they are likely to
entail. That, as already said, can be done without any theory, merely
from the analytic foundations of harmony, as it is generally known.
The theory given above merely shows how these effects are based in
the actual auditory stuff itself.

Of course, only the fringe of this vast subject has been touched.
One of the great difficulties that face the theory of music in general at
the present time is that the art is so highly developed, while—in spite
of the vast amount of analytic work that has been done in connexion
with the form and structure of music—the theory of the basis of music
in the auditory stuff of tones has hitherto been really non-existent.
Reaching the fringe of the subject, therefore, implies much more than
at first appears. We have not yet explained much. For the detail of
that work the centuries remain. But we have at least now a fairly
clear view of the promised land, in search of which men have wandered
so widely and aimlessly.

CHAPTER XX

THE FACTORS THAT MODIFY PARAPHONY

WE may now use the term paraphony to indicate not only the specifically paraphonic intervals of thirds and sixths, but also all intervals— symphonic as well as diaphonic, specially consonant or dissonant,— in so far as they are made, or become, paraphonic, or in so far as their actual or potential paraphonic function is capable of modification by various factors. These factors we have already to some extent encountered in the previous chapters, and shall now proceed to gather together and to resolve as far as possible into their essential functions. We have not only to indicate the effect of each factor, but as far as possible also to explain how that effect is attained.

We notice, then, that paraphony diminishes from its optimum in the thirds and sixths in two directions : in the one towards the symphonies, which create a loss of distinction in unity and thereby a point of relative balance and rest of tonal mass; in the other towards the diaphonies, which make a loss of distinction in confusion and thereby a point of relative restlessness and propulsion of the tonal mass.

We are also familiar with the nature and ground of the varied exposure or ease of distinction of the various pairs of voices : in the following decreasing order,—B–S, B–A and B–T, S–A and S–T, and A–T (cf. above, p. 102 f.).

The effect of an increase in the number of voices is largely due to the mere spreading of the attention. It is more or less a general rule of sensory apprehension that the larger the number of distinct items to be observed simultaneously, the less distinct is each one, and the less easily is it separated from the others and observed, especially in the field of hearing, where fusion and overlapping of volumes play so important a part. The various pairs of voices will, of course, still have the same relative grades of exposure, the outer voices being the most exposed and the innermost ones the least so. But as the maximum grade will now be lower than it was, the minimum will be doubly so; for the larger number of voices makes a larger number of grades. Probably the difference between these grades is also smaller. Thus none of the melodies will be quite so cogent in general, unless one or other is made specially prominent by increased intensity or by specially distinctive or obtrusive tone-blend (as for example in orchestral music). The effect of symphonies or diaphonies upon the flow of melody will also be less marked. Any slight disturbance is hardly noticeable in the mass.

Besides any symphony or diaphony is now likely to be separated by several voices, so that the unity that might otherwise be arrestive will now be variegated, and the clashing that might be confusing will be well spread out and manageable. Even if the thread of two voices is temporarily lost, no great risk will be incurred; for there is much else to engage the attention, and the harmonic patterns of the whole chordal masses will still prevail.

Contrariwise, when the number of parts is decreased, what is allowable in four parts will be governed by apparently stricter rules. The decrease in the number of voices makes each more prominent, so that the rules for it partake more of the stringency of the rules for outer voices. At the same time the innermost voices fall away, and, as the rules for these are the most lax of all, the general average of strictness is apparently raised.

The next question is one of the most important and very puzzling in its own way, unless it is faced with the resolution of rigorous logic : what is the precise nature and basis of the effects of similar and contrary motion?

Oblique motion must, of course not be forgotten. But the answer for it is obvious : one of the voices stands still and is thus already melodically prepared for the attention, so that the other moving voice has the greater freedom and scope so far as the analytic attention is concerned. In other words the melodic spontaneity or force of the moving voice is a matter of its own coherence and expressiveness; and thus the melodically blurring effect of any symphonies and diaphonies that may occur between the two voices is obviated. The early use of oblique motion in connexion with dissonances is familiar. Similarly we have noticed how an otherwise unstable melody has more cogency, and may even progress by leap, if the other tones in the chord remain unchanged, or if the two successive chords differ only in position.

With regard to similar motion, it is exceedingly tempting to interpret it as a positive influence. Its treatment in all textbooks of harmony suggests this strongly. The recent notion of 'exposure' of octaves and fifths (58, 268f.) has greatly increased the force of this suggestion.

But that indication seems to be very misleading. For similar motion has to be kept in relation with all the intervals. And it is important to notice that it has no apparent effect upon paraphonies at all. This would compel us to infer that its effect upon the symphonies and diaphonies is also nought; it merely allows them to stand untouched, relieving neither voice of the necessity for being cogently melodic and paraphonic. This the diaphonies and still more the symphonies fail to be; for while the former urge us on in spite of the confusion they tend to produce, the latter arrest the melodic flow—a most pernicious effect when it is sufficiently 'exposed' by the prominence of the voices that bear it.

The difference between one symphony or diaphony and consecutives is then simply the difference between one bad effect and two of them in succession[1]. One has only to listen to consecutive fifths for a while to notice that the bad effect that appears in them comes out not merely and solely when a second one is played, but that it attaches even to a single one. The fifth,—and to a less extent the fourth,—is a bad interval for polyphony in general, i.e. for paraphony. No doubt it sounds well as a merely momentary or isolated mass of sound, or so long as we think of it as detached from all melodic flow or sequence. But it is commonly recognised to be (otherwise) bare and poor. The octave will, of course, be still worse as an interval than is the fifth, in so far as it is heard as an interval. That happens really only in polyphony. When it is played alone, we tend rather to apprehend it in its (then best) musical function,—merely as a reinforced single tone.

Similar motion, therefore, is a negative condition. The positive force must rather be contrary motion. That will have the effect of favouring in all cases a distinction of the voices,—a distinction both in symphonies and in diaphonies and a greater distinguishability in paraphonies. Contrary motion, in other words, favours the paraphonic effect all round.

The only variant on these conclusions that might be offered is the assumption of a certain degree of blurring and confusing effect in the case of similar motion. The degree, however, would have to be small enough not to affect the thirds and sixths disadvantageously. This modification would still leave the theory on the whole identical with the inferences stated above. In both forms similar motion would rank as a relatively negative factor.

---

[1] This is suggested by Shinn (58, 287) in one instance when he says: "The fact that we admit consecutive fifths when so formed [in connexion with discords], obviously covers the admission of exposed fifths similarly formed."

It may seem to be an enigma in this connexion how it comes that the relations of motion towards a chord that is not yet sounded makes the analysis of that chord easier or harder, seeing that there is no continuity or slur of sound between this chord and the previous one. The objection is well founded. The second chord falls upon us unawares (more or less); many other chords might usually have been played instead. The process, however, is grounded in the nature of melody (in the most general sense) as a motion or phenomenon of motion from one tone to another. A second tone is a sort of reservoir or line of drainage, into which the residual (neural) activity of the first tone runs off and discharges, so that the two become linked together by a line (as it were) of activity which can only emerge when the second tone has made a place for the first to discharge into. Paraphonic differences rest upon the fact that these discharges can be fully controlled only under certain circumstances. And the artistic use of melody requires their full control.

In contrary motion, then, the tones of the second interval both lie ordinally outside or inside the pitch-range of those of the first. The motions between the respective pairs of tones are, therefore, easily distinguishable, and so we get a paraphony—the flowing of two melodies side by side. In the case of successive thirds and sixths this paraphony is quite natural. The tones of these intervals are evidently just the right distance apart for easy melodic flow towards or away from themselves, even though another third or sixth follows. Similar motion does not spoil this effect at all. So in symphonies and diaphonies the bad flow of melody is due to the nature of the intervals, whereby their tones enter into confusion with one another; it is not due to the similarity of motion, at least for the greater part.

We can now see readily the basis of the familiar rules that (melodic) parts should not overlap or cross one another. If they do, there is a great probability of a confusion of melodic connexions and so of faulty paraphony. The same applies to such rules as : "If the same note is found in two consecutive chords, it should in general be kept in the same voice, ... as it will conduce greatly to the smoothness of the part-writing"; and: "Each part should generally go to its nearest note in the following chord" (52, 40)[1]. But if a difference of tone-blend is given, as in choral or chamber music, the voices may move more freely;

---

[1] Cf. a more "modern" expression (51, 14): "There must be a connecting link between successive chords of a note common to each or of one or more parts moving within the interval of a tone or a semitone, the other parts being free in movement."

for the blend of each will serve to bind its component tones together (cf. p. 111, above). It is also apparent why a second should not proceed to a unison; for, without the distinguishing help of contrary motion, this is just confusion worse confounded. In two successive unisons also we tend to lose sight of the intended duality of parts. But one unison is harmless, since the parts must proceed to it from a good paraphony and by contrary motion, while, in leaving it, contrary motion will again be very frequent or an easy paraphony the objective. Thus the two voices drop easily into the one reservoir and as easily discharge towards the following two. The one tone is then really a *unison* psychically. Progression to a unison by similar motion, however, involves the crossing of parts and is only tolerable under favourable circumstances, as by step in one part, or from dominant to tonic, and the like (cf. 52, 31). Relations of visual motion very like these motions of tone have been experimentally established (78).

The interests of harmony are specially concerned in the next fundamental problem : what is the effect of the paraphony of simultaneous intervals upon one another? Or, how do they combine to a resultant? An analogous question emerged naturally in the development of the study of fusion (as the approximation of two tones to the impression made by a single tone), viz.: what degree of fusion appertains to an assemblage of three tones, and how do the individual fusions of its three intervals contribute to the result? But, whatever may be the value of that enquiry in itself, we have already seen that it is not the proper line of approach towards music. The interval is, of course, in a sense the element of structure in music. But our systematic inductions have shown us that the decisive consideration in the function of each interval is how the two melodies of which it forms a phase of conjunction flow together through it. The analysis of music naturally reduces the problem in the first instance to a study of pairs of melodies. For the least grade of harm is the mutual disturbance of two melodies. But there is no reason why three or more melodies should not disturb each other. And music may properly consider these interferences if they occur in typical forms. Our problem is merely an extension of our study from two concurrent melodies to three or more.

But in thus declining the implications of the theory of fusion and in denying its capacity for progress towards a proper theory of music, we do not bind ourselves to reject the serial arrangement of the intervals in respect of something which makes the one preferable to the other (call it 'greater consonance,' if you like) upon which the theory of fusion is based. In that arrangement the major third stands before the minor third and the major sixth before the minor sixth. We shall, however, in due course have to look back upon the theory of fusion or of paraphony or of consonance, or whatever we may call it, and attempt to find a sufficient basis for the functions of which we are now gathering a fuller and completer knowledge.

In approaching this problem of summation, therefore, we must make clear to ourselves first of all what it is that may be added together. In symphony there is a loss of distinction in unity and a consequent arrest of melodic movement. The effect of this loss may, as we have seen, be annulled by the use of contrary motion and other devices. But the unity of the symphony survives to characterise the moment of conjunction of the melodies and to give it an aspect of unity, steadiness, and stability, which must not be heightened if the music is to move smoothly. In the paraphony there is freedom of distinction and of melodic movement. In diaphony there is a loss of distinction in confusion which again may be relieved, as far as approach to it is concerned, by special devices of an analytical tendency. But these of course do not annul the basis of confusion in the stuff of the tones themselves, whatever it may be. They only make the conjunction melodically serviceable and clear in spite of its confusion. Or perhaps it keeps in a state of tension and incompatibility what would otherwise be open confusion.

Thus we see that we can hardly expect to find it possible to add symphonies and diaphonies to a resultant, as we add positive and negative quantities together. But there seems to be no ground of incompatibility between either of these and the neutral paraphonies. And we may safely infer that the conjunction of several symphonies or of several diaphonies will produce an effect of greater unitariness and arrest or of greater tension and harshness.

Thus in *ceg* there are two paraphonies *ce* and *eg* and one symphony *cg*. The general character of the chord is, of course, symphonic, i.e. it is paraphony bound together by symphony. If the c^1 above is also given, there are added a paraphony ec^1 and two symphonies gc^1 and cc^1. The octave displaces the fifth in the outer voices and the fourth appears in a relation of medium obscurity. Thus on the whole the symphonic effect is increased. In the chord bdf^1 there are two (minor third) paraphonies, and one diaphony bf^1, so that the general character of the whole is diaphonic.

But the different exposure given by the different pairs of voices will make differences between chords that consist of the same intervals. We should expect the symphonic (or diaphonic) effect to be greatest when the greater symphony (or diaphony) lies in the outer voices, less when it lies between the soprano and an inner voice than between the latter and the bass[1], and least when it lies in the inner voices. Thus ceg is a greater symphony then $ce\flat g$, because the major third lies in the former on the bass, while in the latter it lies on the highest voice. Of the inversions of the chord bd^1f^1, bd^1f^1 is the worst because the tritone lies in the outer voices, fbd^1 is next because it now lies on the bass, and dfb the best because here it is obscured in the upper voices. The arrangement dbf^1 is even better; for, while the tritone remains in the same place, the minor third now displaces the major sixth in the outer voices, and so increases the tendency towards symphony somewhat[2].

Similarly, of the inversions, ceg is the best, because it contains the two best fusing paraphonies and has the fifth in the outer voices; gc^1e^1, the second inversion, is the next best, and egc^1, the first inversion, the least good, because the latter has not only lesser fusion for each paraphony (3 for III, 6 for VI), but it also has its consonance—the fourth—in the upper voices. In the minor chord the relations are similar for the fourth, but the other two intervals are major in the first inversion and minor in the second. How these sets of changes balance out, is a fine point for experiment to settle. Kemp's experiments show a balance of preferences with respect to the comparative fusions[3] of the two inversions of the minor chord, but a preponderance of preference for the second inversion of the major (29, 207).

Regarding the alternative arrangements for each inversion we may infer, in harmony with musical practice, that, when other things are equal, the closer position is better than the extended one. Thus ceg is better than cge^1, because in the latter the fifth is not in the outer voices, but on the bass. But in ge^1c^2 as against gce^1 we get a minor sixth for a major third, and an eleventh (now in the outer voices) for a fourth. There is the same loosening of relations in ec^1g^1 as against egc^1. But

---

[1] Cf. the rule stated by Külpe and confirmed by Pear's experiments that where the same intervals go to form different chords, the fusional degree of the chord is greatest when the better fusing interval lies lower (29, 209). Cf. 21, Pt 2, § 91: "The more perfect concords ought to be below, and the less perfect above, in a chord."

[2] Cf. 35, 49: "In this inverted form it is classed among the concords of the ancient style."

[3] This is Kemp's term for the object of preference in his experiments. We should not admit the implications of the term in this connexion, of course, as has already been sufficiently indicated.

the differences in these cases cannot be great, since the changes do not produce any bad effects, while the connexions of pattern established between original inversions and their alternative arrangements help to bind them together again. Nevertheless the differences are certainly there, and they call for experimental as well as for theoretical study.

Where diaphonies appear in chords we find in practice that "the lowest fusion contained in the chord is most decisive" (29, 209, 244). That chord is most diaphonic which contains the greatest diaphony. The more exposed it is, of course, the worse will be the effect, as we have already noted and exemplified.

The greater decisiveness of the diaphony appears again in such chords as the dominant seventh. But while the fifth does not outweigh the two discordant intervals so as to render the whole chord purely paraphonic and not at all diaphonic, yet, as we have seen, the discords do annul the symphonic effect of the fifth. The fifth is still a fifth, both as interval and as 'fusion'; but the arresting confusing influence it exerts upon the two streams of melody that run through it, is now annulled by the presence of the diaphonic intervals in the chord, especially if the fifth is not isolated in the bass or soprano. So both one fifth or a second fifth may pass unguarded in the body of such a discord.

The 'root' position gbd^1f^1 is better than $fgbd^1$, because here the fifth rests on the upper voice. In the other two inversions the fifth becomes a fourth. Thus we see that the 'root' positions of chords are not at all due to their derivation from the bass note by any indirect process. The ground of preference lies in the chord itself, not in the supposed partials of one of its notes. If the term 'root position' is to be retained, let it be understood in the sense that in it the chord is, as we hear it, stablest and most nearly symphonic. Moreover, as this stability is referred naturally to the central pitch of the whole chord, i.e. to the bass, the root position of a chord will form a specially good approach to any difficulty in the bass melody, e.g. to a fourth on the bass.

It is a notable fact that both observation and general theory thus place the second inversion of the major chord next to the root position, preferring it to the first inversion. There is no escape from this theoretical conclusion so long as the fourth is ranked as the third grade of consonance. Even Helmholtz, whose theoretical foundations were the successive steps of the harmonic series, was inevitably led to the same result[1] (20, 214). The verdict of experimental observation can be challenged only on the ground of misdirection of description. But it cannot be denied that *a certain* point of view leads to the ranking of the second inversion after the root position and before the first inversion.

But while the musical observations of all time have readily admitted the ranking of the fourth as a consonance immediately after the fifth, the verdict of all polyphonic music has been equally in favour of the inversion in which the fourth is in the upper voices and against the second inversion. In fact the latter conclusion has threatened at times to swamp the other, though it has never really succeeded in doing so completely. The fourth obviously cannot be a dissonance generally, since it is clearly consonant apart from the bass voice. Some special circumstance must be responsible for the bad effect in the bass.

We have already (p. 135 ff.) discussed this question and have noted that its probable basis is the proximity of the (major) third. The fourth somehow suggests this other interval so strongly that it seems itself to be only a point of transition to the third. In thus seeming to call for a resolution, the fourth indeed resembles the dissonances. But the resemblance is only accidental, in the logical sense. We know now that the essential feature of dissonance is not merely its low grade of fusion,— for the sixths have also a low grade; a dissonance shows also a loss of distinction in confusion of the two tones that compose it or of the two melodies that pass through it. For the fourth, however, we cannot claim any lower grade of fusion than that appertaining to the fourth in general; and, even if there is a certain confusion in the bass fourth, it is not an internal confusion that produces merely a loss of distinction, a mere blurring of what is given in the interval; it is a confusion that is due to the attraction of the neighbouring third; the confusion has an external reference beyond the interval actually given.

It is plain to ordinary musical observation that a fourth, exposed not only by standing upon the bass, but rhythmically as well, calls for the third on the same bass, and that, if this call is to be suppressed,

---

[1] Helmholtz referred the effect of the bass fourth to the disturbing effect of tonality. But the tonic does not necessarily come into question at all. On the contrary the six-four chord earliest and oftenest admitted is that of the common chord on the tonic, next is that on the subdominant, and then that on the dominant (35, 86 f.; cf. Prout, p. 133f., above). On the other degrees of the scale it is rarely used. Thus one might even claim that, far from being disturbed, the chord is perhaps rather made tolerable by the influence of a distinctly tonic reference. This would be confirmed by the fact that the peculiar character of the six-four chord was recognised long before tonality had come clearly to the surface of the musical consciousness. In fact, its character was then more stringently unique than later.

not only must the rhythmical exposure[1] be avoided, but a cogent melodic line must also be driven through the bass note. And at least a great part of the necessity for also approaching a bass fourth in a special way is likewise due to this proximity of the major third. If the question be raised how this third can possibly disturb the melodic flow of the bass when even the fourth itself is not yet sounded, we must answer by drawing attention again to the fact that a melodic point that is about to follow upon one just sounded, provides a sort of outlet for the latter's residual energy. If the second point lies near the first, the discharge is easy and cogent. The further apart the two points lie, the less forceful is the transference. A melodic leap is in itself a difficulty. But this difficulty can be increased by various circumstances : by the unfamiliarity of the leap to be taken, and especially by the proximity of an easier—more consonant, more paraphonic, or more familiar—interval. The difficulty of the fourth from the bass is probably due to this latter circumstance. The fourth is more consonant, it is true; but the (major) third—as the most consonant (i.e. best fused) paraphony —is both very important and very interesting. And the intonation of a bass fourth brings any slender (leaping) melodic line so close to this attractive third that the actual fourth will come as a sort of jar upon the expectation. The melodic flow will be disturbed. Hence the necessity of making the melodic line leading to the bass note cogent by the devices summarised above (p. 134 f.).

When the bass note is firmly established, the third suggested is that upon it. When it is not well established, the third suggested must be that below the upper note of the fourth. This is strongly indicated by the effort all the rules make to improve the cogency of the bass melody. As the bass is the most exposed voice of all, any vacillation in its course will make the whole chord unstable.

But there can be no such instability when the fourth is not on the bass. When, as in *cegc*[1], it lies in the upper voices, all the rest of the

---

[1] Rhythmical exposure is of the greatest importance in music generally in so far as it increases or decreases the good or bad, desired or undesired, feature of any purely tonal element. The arrestive effect of the last chord of a cadence, for example, is heightened by such exposure, as it must be if the 'cadential' effect is to be attained. It is at such points that one sees the continuity between the 'dynamics' of tone and of rhythm—a large subject which deserves special treatment. But that can only be given when we have brought the dynamics of tone into acceptable order. The dynamics of rhythm are naturally more obvious and apparent. Among the ancient Greeks the dynamic of arrest was induced largely by lowness of pitch. But the octave also appeared as a concluding symphony (16, Index sub. 'Symphone,' 'Grave,' and 'Aigu'). Both of these effects are still generally valid.

chord, including its bass, calls for c^1, which is at the same line a great consonance, whereas b would give a great dissonance. Similarly in egc^1 the two paraphonies exposed upon the bass make c^1 easy, while b would itself as leading note suggest the c^1 above it.

This course of reasoning strongly suggests that the basis of the bass fourth anomaly must be merely one of a class of similar cases. Now the anomaly of the fourth is due to its tendency towards the neighbouring third. Hence we might express the principle of a class of such cases as due to the proximity to one another in size of intervals of different fusional or paraphonic character, and to a resulting tendency of the one to point to, or to fall into, the other.

Methods of 'ear-training' make use of this feature of intervals as well as of others. Thus the scheme propounded by Bridge and Sawyer (3, 194) is as follows : Octave—perfect unison of the two sounds so that they sound almost as one; VII—upper note will require to ascend; 7—both notes will require to move, upper down one degree, lower up a fourth; VI—upper will tend to fall to the dominant; 6—very strong tendency to the dominant; 5—sounds bright, ear at rest; diminished 5—both notes will require to move towards one another by a second; augmented 4—both will require to move away from one another by a second; 4—upper requires to fall to the mediant; III—calm, peaceful effect, ear at rest; 3—peaceful, but more melancholy, ear at rest; II—close discord, lower note requires to descend a semitone; 2—a fierce discord, lower note requires to descend a whole tone.

These rules for the recognition of intervals make use of a variety of aspects : (a) symphony—octave and fifth; (b) paraphonic excellence—III, 3; (c) generalised 'harmonic' (i.e. polyphonic or paraphonic) disposition—7, diminished 5, augmented 4, II, and 2; (d) tendency towards proximal interval of highly symphonic or paraphonic nature—VII to octave, VI and, still more, 6 to fifth, 4 to major third. Other distributions of these aspects seem to be possible, for example : (a) symphony in its grade—0, 5, 4; (b) paraphonic excellence or repose—III (small and bright), 3 (small and sad), VI (large and bright), 6 (large and sad, or—as discord—tending to fall to fifth); (c) dissonance and size—smallest 2, next II, largest VII, next 7, middle size the augmented 4 and diminished 5; (d) proximity to a marked consonance—VII to octave, 2 to unison; second degree of proximity—7 and II; (e) generalised 'harmonic' tendency—as familiar. Or other rules might be formulated according to the taste and fancy of the recogniser.

Similarly Bridge and Sawyer's rules for the recognition of single tones, which presuppose a given (and retained) tonic, are : tonic (and octave)—at rest; leading tone—up; submediant—to the dominant; dominant—bright, bare, satisfying to the ear; subdominant—dull, to the mediant; mediant—a calm, peaceful sound, on which the ear may rest; supertonic—to the tonic.

We do not feel any surprise at the major seventh's reminding us of the octave nor of the minor second's suggestion of unison. In each case the other interval is nearly there, as far as the volumic proportions are concerned, and is a very characteristic thing. Apart from the tendency of the fourth to the major third, what is most notable is the connexion established between the major sixth and, in a stronger degree, between the minor sixth, and the fifth. We might have expected both these intervals to share the repose characteristic of the thirds. The major sixth undoubtedly does so markedly; and it is debatable whether it does not really do so much more than it points to the fifth. But in the minor sixth this latter tendency is indisputable. Here, then, we have another example to the fourth—a paraphonic-consonant interval with a tendency of 'resolution.'

In fact, we have more—an exaggeration of the feature of dissonance often ascribed to the fourth : for the minor sixth—at least in equal temperament—is an undoubted dissonance whenever it functions as an augmented fifth. It is so even when the chord it appears in otherwise contains no dissonance at all—$ceg\sharp$ (two major thirds). When we substitute $e\flat$ or f for e, we get $(ce\flat a\flat)$ the first inversion of the common chord on the major tonic $a\flat$ ($= g\sharp$ in equal temperament) or the second inversion of the minor common chord on f $(cfa\flat)$. Both of these chords are generally consonant, apart from the bass fourth in the latter. In particular the minor sixth is now consonant, although in the six-four chord it still shows a tendency to fall to the g (the 'dominant,' as it were, of itself and of the rules quoted above).

Of course the problem of the dissonance of $c(e)g\sharp$ is not solved by a simple reference to the beating of partials. For although the third partial of c ($= g^1$) and the second partial of $g\sharp$ ($= g^1\sharp$) beat with one another—upon a c of 144 vibrations per second—18 times a second, *if these partials occur in the given primary tones*, we must not forget that the fourth partial of c ($= c^2$) and the third partial of e ($= b^1$) beat 36 times a second, or, if the interval is lowered an octave, 18 times a second. If the major third is not a dissonance on 144 vibrations, it cannot be a consonance on 72 vbs. if $cg\sharp$ is a dissonance on 144 vibrations because of its 18 beats per second. The same argument applies to the interval *cab*, which would give 27 beats a second. We must, therefore, look for some other difference than that of a few beats per second. After all, the dissonance of a major second has only some 25 beats (between the primary tones) in this region, so that it should sound better than the minor sixth just mentioned, unless we admit that the beats of partials are of less effect than those of primaries. But why should they be so? And if they are, and if *cg♯* has no partials, or at least not those that make 18 beats per second (only one of the beating partials need be absent!), then *cg♯* would be consonant. But we cannot revive the problem of partials as a whole again here.

In both the chords *cebab* and *cfab* we have a grouping of a minor third, a fourth, and a minor sixth. They differ only in their distribution. In *ceg♯* we have two major thirds and (in equal temperament) a minor sixth. If we argue that in the latter chord the outer interval is heard as an augmented fifth, we must ask : why? It is certainly often claimed that the ear has the wonderful faculty of hearing what a chord should be ideally instead of what it really is. Equal temperament is commonly said to rest upon this basis. But how does the ear know what should be? Why does the displacement of the middle tone by a semitone upwards or downwards incline it to hear *cg♯* (= *ab*) rather as a minor sixth and consonant than as an augmented fifth and dissonant? Surely not the pitch of the middle tone, but the intervals it forms with the outer tones. Then a minor third and a fourth somehow call for a minor sixth, while two major thirds call for an augmented fifth. But why should not two major thirds in the first instance have called for *cg♯* to be a consonance as well as *cab*? If answer be made that *g♯* is not a tone of the original scale, why, we must ask, did not this relation of it to two major thirds call for its introduction into the scale, as other chordal connexions have called for changes in scales (e.g. the sharp leading tone in the minor scale)?

The question is certainly difficult. It would be foolish to try to make it look easy. Probably the best reason is the one that appears first in the musical consciousness : that *ceg♯*, because of its major third in the bass and the approximation of the two other intervals to the corresponding ones of the major triad reminds us so strongly of this triad that we feel the discrepancy as a distinct jar, confusion, or dissonance, just as we may feel a certain approximation to the portrayal of a familiar face as a caricature or as an artistic offence. This feeling can only be supported by the fact that the interval in question—even as a minor sixth—stands very low in the scale of fusions. When the minor sixth stands alone, mere proximity to the fifth would make it suggest the latter very strongly and so appear dissonant. Its paraphonic capacity would not yet have been exposed by its subdivision into a minor third and a fourth. So we find a prevailing tendency in early music to class the minor sixth as a dissonance (cf. 60, 2).

Thus we come upon the general problem of the natural tendencies of intervals to suggest one another. Of this the problem of the resolution of dissonances forms a sub-division.

Probably the chief consideration in the latter problem must be the cogency of the melodic movement of the voices[1] that constitute the dissonant interval—an interval that impels melodic movement forwards and so calls for 'resolution.' Thus, for example, the tritone is commonly said to resolve generally as a diminished fifth inwards to a major third and as an augmented fourth outwards to a minor sixth. If, on the basis of this generalisation, we were to claim, or to attempt to establish by experimental means, some special connexion between these intervals, we should not only find ourselves greatly embarrassed by the existence of other forms of resolution of the tritone, but we should find it hard to justify the distinction between the two forms of the interval—diminished fifth and augmented fourth,—and their so different resolutions. A much easier and more natural solution would be to claim that the tritone must generally 'resolve' by the most cogent motion of the voices it carries on; and, apart from other modifying circumstances, the smallest step is the most cogent. Hence the best resolution is by the movement of a semitone in each voice (cf. above, p. 172, note). And, unless we are to leave the key of the chord to be resolved, this procedure would give the commonest resolutions of the two forms of the tritone mentioned. We should by no means thereby be precluded from making other resolutions of a chord containing the tritone, such as by movement of only the lower voice of the tritone by a semitone, or by different movements of the other elements of the chord than those of the tritone. We should expect the 'harmonic' value of all possible resolutions, i.e. their frequency in the great masters or their beauty, to vary directly with the melodic cogency of the different voices of the chord, not

[1] Cf. D. F. Tovey's interesting remark: "Even the modern researches of Helmholtz fail to represent classical and modern harmony, in so far as the phenomena of beats are quite independent of the contrapuntal nature of concord and discord, which depends upon the melodic intelligibility of the motion of the parts" (74).

forgetting, of course, that the cogency of a movement, though primarily determined by its shortness, may be increased by other factors, such as consonance (step of fourth, fifth, or octave), tonality (as in the leading note), an established figure or 'sequence,' the symphonic coherence of one or even two otherwise undetermined movements with the specially motivated and cogent movements of the other voices, etc.

The range of ordinary variations may be exemplified more fully with the interval of the minor seventh on the dominant in the keys of *c* major and minor. It is obviously not here a case of the mutual suggestiveness of intervals, but of the melodic movements of voices resulting in different intervals which are then accepted in so far as they are consonant or paraphonic or conform with the intended tonality or are possible companions for the tones resulting from the other voices, and so on.

Nor do we need to assume the existence of a mutual suggestiveness between whole *chords* as such (or between 'harmonies'). Any such connexions or suggestions that may occur and be felt are merely a scheme or shorthand of melodic connexions, having as a scheme no binding force or compulsoriness at all. That belongs essentially to the melodic transitions in detail and in their mutual compatibility, as already indicated. Thus in the generalised connexion between the chord of the dominant seventh and that of the tonic there is clearly a parallelism of the patterns that result from the combined melodic transitions or resolution. Neighbouring tones of each chord lifted up an octave retain their melodic connexions unchanged.

In the case of the connexion of chords by tonality, we reach a new level of melodic grammar, so to speak. Instead of constructing from single words we make use of familiar phrases. The connexion between dominant or subdominant and tonic or *vice versa* is so familiar and their chords so characteristic that we come to know exactly what to expect. Thus the melodic connexion establishes itself more easily. Besides,—and perhaps this is the root of the matter,—in these progressions the very symphonic effect produced may be just what is desired at that point of the music, namely, a tendency towards arrest or close.

The existence of an effect due to the connexion of chords a third apart has been claimed by Shinn, who believes in a grading of effect in three steps, corresponding to fifth-fourth, sixth-third and seventh-second intervals between the chords. To some extent Shinn fails to distinguish these connexions from the intervals between the pairs of voices in question. The establishment of this grading would require an extended statistical foundation, chosen with due regard to the elimination of other differences. The problem raised by Shinn is of considerable importance, but the factual and deductive grounds for his generalisation seem to be at present insufficient.

It has been claimed occasionally that downward motion favours melodic continuity, and so tends to annul unparaphonic effects. Here again we lack a proper basis from which to confirm or to oppose this suggestion. We know that descending intervals are found harder to judge at first than are ascending ones. Whether the intervals of melodies descend oftener than they ascend has yet to be determined (cf. 16, 173 ff.). The effect of a progression of roots of chords by leap of a third is said to be generally more satisfactory downwards than upwards, especially if two or three such leaps occur in succession (52, 56; cf. 60, 35).

CHAPTER XXI

RETROSPECT AND THE OUTLOOK FOR THEORY

THE division of intervals into the three classes of symphony, paraphony, and diaphony, which we have attained and expounded in the previous chapters, stands in opposition to the progressive grading of intervals derived from the studies of Helmholtz and Stumpf. In these the octave forms an extreme fusion of apparent unity in simultaneity or smoothness in transition between tones, from which there is a gradual variation towards complete two-ness or harsh beating in simultaneity or complete difference in succession. We have reflected briefly from time to time upon this change of theory, but it seems necessary to consider it more explicitly now.

Helmholtz's quantitative estimation of grades of consonance between simultaneous tones was based upon mathematical calculations regarding the roughness of beating between their component partials (20, 192 ff.). In so far as a connexion between successive tones might be established by their common partials a certain smoothness of transition could be achieved without which tones would be simply different; but there could be no harshness or dissonance. In the case of simultaneous tones beating would also appear so far as the partials presupposed actually existed in the tones; and the roughness felt in any interval would vary with the partials contained in its tones. But in so far as the partials might lapse without alteration of the grading of intervals according to consonance and dissonance, the theory would be fatally discredited. No force of memory could save it. It is necessary to recall these points from a previous chapter because they compel us to face the task of constructing a theory of consonance and dissonance without the help of coincident or beating partials, except it may be as a secondary or supporting factor making for confusion or loss of distinction.

Stumpf's grading is based upon direct experimental observations confirmed and amplified by the work of others. His description of fusion, especially in its highest grades, as approximation to the unity of a single tone, is a true representation of the experimental judgments and corresponds fully with the descriptive definition of 'symphony' given by the Greeks, which is thus seen to be correct both as far as it goes and inasmuch as it does not venture farther. We have adopted it in our exposition above, and have found a sufficient and satisfactory parallel to it in the volumic coincidence of tones, at least for the cases of the octave and the fifth. These yield a volumic mass that in two distinct grades resembles the symmetrical balanced mass of sound that constitutes the pure tone. There is no doubt that this part of the theory must be maintained in all future theories of consonance and dissonance. It fully explains the symphony of the octave and the fifth.

But in the other two divisions of paraphonic and diaphonic intervals our scheme fails to coincide with Stumpf's. It is only in so far as we can look upon intervals as mere static and isolated masses of sound that we can attempt to force the aspect of 'approximation to unity' down the whole series from octave to major seventh. But this static aspect is practically excluded from music and is hard to realise experimentally unless with unmusical persons or with neglect of analysis (cf. Kemp, 29)[1]. The new description is one that establishes a plain continuity between intervals (music in two voices) and music in any number of parts. In paraphony, we have shown, there is an equal distinguishability of the tones of an interval, so that two melodies may run side by side through it without any mutual interference. In diaphony this confusion re-appears, although the tones are not now apparently one, but, as we can readily infer from their harsh coincidence, obviously more-than-one-like. The major seventh and the minor second strongly declare their dissonance. But they do not necessarily limit themselves to a semblance of *two*-ness, as the erroneous judgments of untrained and unmusical ears often indicate (64, 371 ff., *et passim*). We have to find a sufficient basis for this positive diaphony (or dissonance, which in music is always felt as a positive harshness not sufficiently described by mere two-ness or difference[2]) in the stuff of tones, and—to emphasise it again—without the help of beating partials. Even beating difference-tones must be kept in reserve for any special purpose they may properly be able to fulfil[3].

---

[1] Kemp and Külpe go much farther than Stumpf, who refuses to see any change in a fusion owing to the presence of other tones.

[2] A much better description is, for example, this: "discordantia est duorum sonorum sibimet permixtorum dura collisio" (*Quisdam Aristoteles*, 6, 260).

[3] A claim to the recognition of neutral grades of sonance was made by F. Krueger in his lengthy discussion of Stumpf's criticism of his difference-tone theory of consonance and dissonance. Krueger pointed out that the psychological relativity of the notions of consonance and dissonance and the historical mobility of the border between them was noted by Helmholtz. We have already seen how little real truth there is in this historical appeal. Stumpf was also tempted to form a class of neutral intervals,—the sevens in particular,—but was clearly aware that his notion of fusion could give no such

Now the whole course of our new theory of tones advises us to look around the main points of our actual conclusions for lines of development rather than away from them to outlying regions or to adventitious phenomena. Our grading gives three regions with special points in each, namely :

Symphony.	Paraphony.	Diaphony.
Loss of distinction in unitariness	Ease and equality of distinction	Loss of distinction in confusion
p, o, 5, 4	III, 3, VI, 6	T, 7, II, 2, VII

(We have added the prime to the head of the series as our study of its functions in music has justified our doing so. The prime, as the older writers on music so often said[1], and as the modern (acoustical) theory of music has always failed to understand or to explain[2], is the greatest consonance of all.)

opposition as is implicit in the concepts of consonance and dissonance. Krueger's theory does not really justify his formation of a neutral class either, however true it may be that various intervals might well be placed in either class of consonance or dissonance. And it is difficult to see how such an appeal to a neutral class can save a theory from special criticism after it has been shown that such very different theorists as Helmholtz and Stumpf also noted the existence of intervals of doubtful or indifferent character and yet equally failed to justify the formation of a neutral class. Krueger believed that an interval tended towards neutrality of sonance in so far as it was sounded more gently or briefly, or was raised towards the higher end of the musical range of pitch, or was extended by one or more octaves, etc. (v. 31, *passim*). Neutrality is given when "for any reasons the consonantal or dissonantal characteristics postulated by the theory are lacking or reach a certain limit of clearness or of effectiveness in the whole sound, or, again, when one and the same chord contains both consonantal and dissonantal characteristics, and these two sets, not greatly marked, annul one another in the total impression" (32, 314). Consonance is due to the coincidence of difference-tones, of which Krueger asserts the existence of five or six (theoretically) for each interval; dissonance to the 'neighbourly interference' of two or more of these five or six, i.e. to their beating and indistinguishability, etc. Thus we get an approximation towards unity and indistinguishability both in the greater dissonances and in the greater consonances. Why two primary tones and two loud difference-tones should suggest more oneness than these two primaries with five difference-tones that are proportionately weaker is not explained: especially when we recall that the difference-tones need not, and in the vast majority of cases are not distinguished at all. Surely the fewer more distinct elements should be 'rougher' than the fainter ones. Besides, Stumpf's criticism of the facts regarding these five or six difference-tones robs the theory of its basis (cf. 68, 69 and 70).

[1] E.g. Johannes de Garlandia (6, 104): "Concordantia dicitur esse quando due voces junguntur in eodem tempore, ita quod una potest compati cum alia, secundum auditum... Perfecta dicitur quando due voces junguntur in eodem tempore, ita quod una, secundum auditum, non percipitur ab alia propter concordantiam, et dicitur equisonantiam, ut in unisono et diapason." Unison, as already said, has only meaning for polyphony; for only then can one sound be held and heard to be the conjunction of two.

[2] Cf. 64, 178: "the hypothetical fusion of the prime."

In order that we may look around these intervals for some means of progress, let us arrange them according to their place in the ordinal series of an octave. Then we get :

Symph.	Diaph.	Paraph.	Diaph.	Symph.	Diaph.
p	2, II	3, III	?	4	T
o	VII, 7	VI, 6	6	5	T

The two ends of the series are constituted by the greatest symphonies; next them stand the greatest diaphonies; a lesser diaphony leads through the two grades of paraphony to a middle region in which stand the consonances of the fourth and fifth, separated by the diaphonic tritone. This arrangement suggests a regular or periodic change from the extremes towards the middle region, in which range the diatonic intervals mark out familiar points.

But, we must ask, do these intervals really designate *points* in the range of intervals included within the octave? If both the minor and the major second are diaphonies, and both the minor and the major third are paraphonies, should we not rather speak in general of a diaphonic or paraphonic *region*? And should we not also in particular speak of a region having, for example, the paraphonic degree of the minor third? (Remember again that we have lost the treacherous aid of beating partials!)

Arguments of some strength may be urged in support of such a view, startling and even outrageous as it may at the first glance seem to be[1]. These rest upon facts that are fully familiar to students of

---

[1] One can see, however, that Stumpf felt himself drawn towards this conclusion (cf. 64, 176 ff.): "Doubts about the sinking of the curve between [major and minor thirds, etc.] have another special cause in that these intervals differ only by a semitone and so the intervening are always apprehended as a deviation from them, as a mistuned third etc." So, also, Helmholtz, who wrote (15, 200): "two simple tones making various intervals adjacent to the major third and sounded together will produce a uniform uninterrupted mass of sound, without any break in their harmoniousness, provided they do not approach a Second too closely on the one hand or a Fourth on the other. My own experiments with stopped organ pipes justify me in asserting that however much this conclusion is opposed to musical dogmas, it is borne out by the fact, provided that really simple tones are used for the purpose." The later experimental work of Stumpf and Meyer (60) has disproved this in so far as the ability to recognise the interval with great approximation to its pure form is concerned. (Helmholtz went so far as to say it was impossible to *tune* perfect major or minor thirds on stopped organ pipes or tuning-forks without the aid of other intervals.) Stumpf and Meyer's results prove the ability to find the almost exact interval even with *successive* tones. Here there is presumably no 'fusion' of any kind, beats, difference-tones, or 'fusion' proper,—so that we are all the more thrown back upon an accurate sense of interval (to which we have given a positive basis in our theory in volumic proportions, and) which is presumably most developed in the most highly

hearing and of music, but that have been rather coldly entertained by those who have aspired to construct a theory of the foundations of music. This neglect is, of course, natural and inevitable so long as the facts in question seem to be not only incompatible with the prevalent theoretical point of view, but even to be artificial and arbitrary distinctions or classifications unattainable from the straight roads of theory.

One argument is familiar to all those who possess absolute ear and to others from their records. Such a person does not, for example, hear the progressive transition from a good d to a good e as a sharp lapse of 'd' into 'something not d, but higher than d,' then '$d\sharp$,' then 'flatter than e,' and finally 'e,' or the like; but as a region of 'd colour' surrounding the 'best d,' bordering on a region of $d\sharp$ including a smaller optimal region, passing into a new region of e leading to a best e, and so on. Each tone name characterises not one rate of vibration or a very narrow zone defined by the range of hardly distinguishable differences, but a relatively large zone of perfectly obvious differences. An absolute ear also adjusts itself readily enough to differences in the pitch to which a musical instrument, such as the piano, may be tuned, if the difference is not great enough to carry a tone into the zone of its chromatic neighbour. In this case keys are interchanged, c major becoming $d\flat$ major or b major. In some cases of very rigid absolute ear, however, even smallish differences of tuning are disturbing.

Another is the similar usage so characteristic of music of applying the same pitch names to zones of pitch. Thus, e.g. we have not only $d\flat$, d and $d\sharp$ but also $d\flat\flat$ and $d\times$ ($\sharp\sharp$)[1], terms which are acoustically so anomalous, as they are on instruments of equal temperament exactly, and in any case very nearly, equal to the pitches of c and e. Similarly the names of the intervals are not only the same for their major and minor forms, but they are extended to include augmented and diminished forms. An augmented second, for example, is more or less the same as a minor third as far as pitch is concerned; but, as music has always claimed, the augmented second partakes of the diaphony of the major second, not of the paraphony of the minor third. And an augmented

---

trained observers such as Stumpf and Meyer had. Hence in observers such as Helmholtz, who do not possess this faculty, we may assume we get an approach to the effect of the consonantal value of the tones apart from their intervallic precision and apart from difference-tone values which Helmholtz obviously failed to make use of in these cases. Hence a neutral (or as I call it, a paraphonic) range from second to fourth or thereby.

[1] The original points for these are, of course, not d, but $d\flat$ and $d\sharp$ respectively, so that there is a second grade of flattening or sharpening here only in a terminological sense.

or diminished interval, when it occurs in full force, as between $a\flat$ and b in a minor scale is found to be peculiarly hard to sing. When the same transition of pitch occurs in a true minor third, as from c to $e\flat$ in c minor, no such difficulty is felt.

We may recall a quotation from Stumpf (p. 144, above) : "That one and the same unmodified pair of tones should now fuse more and now less according as we apprehend it as c–$e\flat$ or as c–$d\sharp$ is out of the question, because fusion is a function of the two sensations—or of their physio-logical bases—and can change only with these same" (71, 328). This utterance implies the very sharp distinctions of the points of the scale and the fusions peculiar to them that we have already mentioned. But, if pitches and degrees of paraphony belong, not to points optimally and only, but also in some degree to zones or regions, there seems to be no insuperable barrier to our extending these zones for various reasons and to ascribing to a tone the pitch name, or to an interval the paraphony, of the zone within which it is included. Then in con-nexion with these musical reasons of key relationship, modulation, etc., we shall apprehend a tone of a certain number of vibrations as a modified d, i.e. as $d\sharp$, not as $e\flat$, with which it may more or less coincide, and we shall then attach to it the pitch name of the (extended) region we ascribe it to and the paraphony of that region, namely the diaphony of the seconds. But its diaphony will not be greater than that of the major second as is the minor second's, but less, of course; still, however, diaphony, and not the paraphony of the third. The paraphony of an interval will be determined primarily and chiefly by its precise nature as an interval, but within any zone much will also depend upon the circumstances of our approach to it in the flow of the music.

Reverting to our series within the octave we may now note further that the great symphonies are bounded by the strongest diaphonies. In the former there is a special fit or balance of the tones; in the latter we have the worst coincidences of volume, as soon as the range of hardly perceptible difference is passed. The zone of symphony in the octave and fifth is very narrow, allowing of no degrees, no major or minor forms. In the diaphonies bordering on the symphonies the pitch centres of the tones either encroach upon one another (small intervals) and so make each other hardly distinguishable, or the lower end of the volume of the upper tone and the pitch centre of the lower fall close together (large intervals), making a notable blemish in the centre of the lower tone, and, in the major seventh, an approximation to the octave which it thus strongly suggests. We probably do not shift our point of attention in these two groups from one side to another of the pitch of the lower tone. We hear each tone in a mass as a whole, including its pitch and the length of its volume; then its paraphonic nature will be decided by whatever excessive proximity of parts or relative freedom and independence of them there is on the whole.

In the paraphonies we reach a zone of volumic proportions in which the pitches of the tones stand well apart, are therefore easily discriminable without mutual interference, and in series can readily form distinct melodic lines, not running together or blurring one another.

Round the fifth again we find a border of diaphony in the tritone and in the minor sixth when it appears as an augmented fifth or when it is, as in early music, not marked off by its place in chords as a paraphonic sixth. This fluidity of character confirms the extensibility of zones we have already noted. And it also seems a sufficient explanation of the fact that in the sixths the lesser paraphony follows upon the greater, not conversely as in the thirds. If the fifth stood not where we find it, the interval next beyond the major sixth (i.e. the minor sixth) would presumably be a better paraphony than it actually is.

The only point that remains for discussion concerns the fourth, that old bone of contention. Our scheme tempts us anew to consider it a dissonance, less diaphonic than is the tritone. So one might plausibly account for its having been so often termed a dissonance. But even so it might at the same time be held, like the sixth, to be includible in the paraphonic zone of the thirds, being then a little more paraphonic than the major third. But, on the other hand, we cannot get round the fact that the fourth is a consonance of third, though low, degree. To what extent its being the inversion of the fifth is responsible for this, it might be difficult to decide. The connexion felt between its symphony and its position in the upper voices of chords would support any doubt as to its being in itself symphonic, as would also the large (statistical) gap between the fifth and the fourth and the small gap between the latter and the thirds (cf. 77, 58). Our failure to find a form of distinctive volumic balance for the fourth would tend towards the same conclusion. So we should infer that the fourth is a paraphony or even a diaphony whose cousinship to the fifth has given it the rank of symphony. This conglomeration of relations would suit the unstable character of the fourth in music. An incautious fourth from the bass would then be a sort of augmented third or a diaphony of lesser degree than the tritone. It would have the fluid nature of the minor sixth between diaphony and paraphony, sometimes 'resolving' therefore into an ordinary (major) third. There is in the relation of inversion as such no reason for any transference of symphony from the fifth to the fourth. But when the fourth occurs as the top of a major common chord, it is then the complement to an actual fifth.

But if the range of the octave is characterised by these wide regions of paraphony and diaphony, what, the reader may ask, accounts for the very precise form of our present diatonic scale and for the perfection that so many claim for just temperament. Here we return to the problem of the scale, which is primarily the problem of the skeleton of divisions of the octave that will at any time offer the greatest and most convenient scheme of intervals and chords, etc. The octave and fifth are inevitable land-marks of great precision and force. The fifth may be taken upwards or downwards, yielding $cfgc^1$. And from the whole tone thus derived we may get various farther divisions of which the chief type attempts to subdivide the two large intervals—the two 'tetrachords.' It is a familiar fact that many forms of subdivision have been tried and tolerated as more or less permanent modes. We ourselves have two—the major and minor, the latter of which the theories that have worked with harmonic partials have never been able to justify. And in bagpipe music we find a neutral third midway between our major and minor sections. The extensibility of zones, it may be noted in passing, is richly confirmed in the familiar tendency of the ear trained upon our two modes to hear the intervals of exotic music as approximations to the intervals from the same zone that occur in our music.

Of course the burning question is: why has our harmonic style so favoured our two sections in the precise form they have taken, and especially the major one? Here the appeal to the correlation with the ratios of harmonies seems inevitable. To deny may seem like wilful scorn of a providence governing musical theory. But at this point difference-tones seem to be much more important than upper partials. For the latter are not only accidental and distant, but as Macfarren said, they should really create quite a babel of confusion; whereas the loudest difference tones of 4, 5, and 6 would necessarily be 1 (twice), 2 (twice), 3, and 4; or with the primaries 1 (loud), 2 (loud), 3 (not weak), 4, 5, and 6. Here we have two octaves 1–2 and 2–4, and two fifths 2–3 and 4–6. A deviation from 5 in either direction would upset the two octaves and one of the fifths, thus creating in the whole sound a somewhat indefinitely located, but quite noticeable, harshness. The minor third, on the other hand, is sufficiently justified by its being the remainder of the major third from the fifth (which two would between

them yield all the tones of the major diatonic scale). Any scale determined by these two intervals would also yield minor third chords,
which would thus become familiar parts of the system, tolerated as
such, in spite of their less symphonic resultant, which is familiar to every
musician.

Two points may be emphasised with reference to the above. First,
we are not here attempting to justify the thirds *ab origine* by this means;
they carry their own justification as paraphonies within them; the only
question is: why has our harmonic music so favoured the particular
points of the paraphonic zone at which the major and minor thirds
stand? Second, we do not forget that the major mode was by no means
the fount and origin of all music; it is rather the culminating result
of a long development. But it stands in modern music as the logically
prior skeleton of our system, and as such it requires special justification.

It is open to the musician, of course, to make many other sections
of the octave than our two, and to build a perfectly coherent music
of somewhat limited scope upon them. This may be forcibly extended
even some way into the range of polyphony. But the fascination of
the great consonances gradually drags it into the lines of the two chief
steps of paraphony and the great systems they yield.

With this free outlook we seem able to do justice to all kinds of
music without the ridiculous restrictions and sophistries of the harmonic
theories. Of course, all the theories try to bring themselves into line
with music so as to explain its ways; and in so far as theories succeed
in doing so, we may be all taken as in agreement with one another.
The only question is: whose basis of explanation is the primary one?
And we may claim that honour for ours for two reasons: (1) the basis
of explanation is inherent in the primary tones concerned in the
phenomena to be explained; (2) the new facts we have built upon have
not been derived from sources external to music so as now to be
problematically read into music, but they have been derived from the
empirical generalisations of music itself.

The explanation suggested for the nature of the intervals between
the fifth and the octave would obviously hold equally for the intervals
greater than the octave up to the twelfth. For in the ninths the ends
of the upper volume will fall near the pitch centre of the lower tone
and will so mar its outline, while in the tenths and eleventh these
points will draw apart so that they will be readily distinguishable and
independent. In the twelfth we meet again (cf. 77, 68, 111f.) with a
sort of balance or symmetry of volumic parts, the upper half of the

volume of the lower tone being divided into three equal parts by the
pitch and the lower end of the upper tone's volume. So we may well
have diaphonies in the region on either side of this point. But there
seems to be no further ground for establishing paraphonies, such as
the compound sixths. Perhaps the diaphonies around the double
octave might pass as blemishes on the division (4, 2, 1, 1) peculiar to
the latter.

At this point, feeling we had pushed the range of the direct and
inherent distinctions of intervals well out beyond the octave without

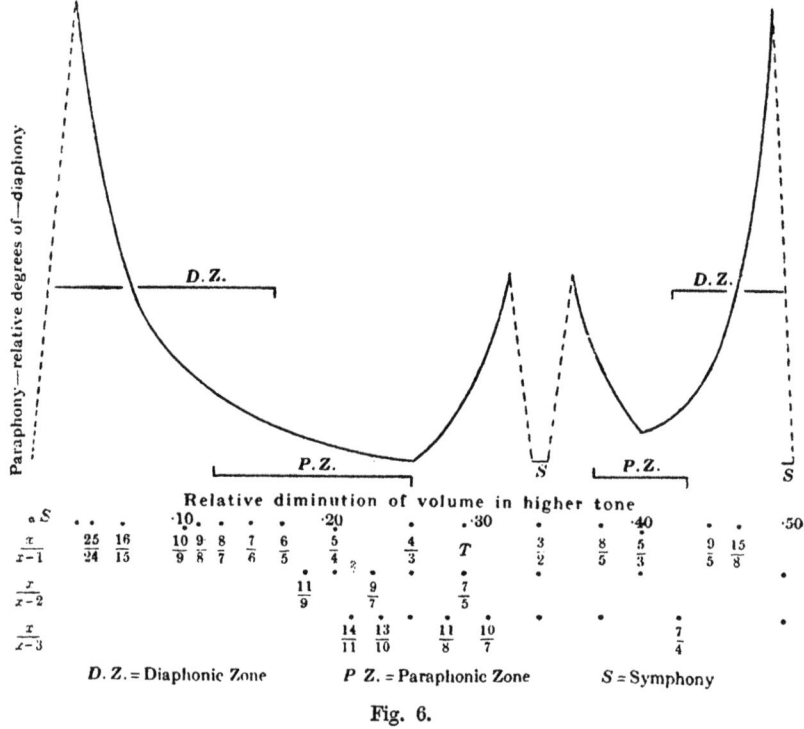

Fig. 6.

appealing to any extraneous or secondary factors, we might perhaps
venture to ask the functions of memory to support us or invite the
aid of harmonics. These functions must certainly accrue at some time or
other; for there can be no doubt that we come in time to be thoroughly
familiar with proportions and divisions of tonal volume by mere memory.
As we have seen, all our sense of interval is founded upon that ability.
But we must take all care to justify our musical distinctions by powers
that dwell in the stuff of tones as they are actually presented to us,

before we appeal to any outlying circumstances or to habit. We must bring our engine into motion with its own steam, ere we can expect to steady its action by its acquired momentum.

The results of the above suggestions may be embodied in graphic form (fig. 6). The height of the curve represents in some fashion the comparative diaphony or paraphony or symphony of the intervals below it. The height has been taken in most cases as the denominator of the fraction by which the volume of the higher tone is less than the lower, thus $\frac{1}{30}$ or thereby (for a quarter tone), $\frac{1}{16}$, $\frac{1}{9}$, $\frac{1}{6}$, $\frac{1}{5}$, $\frac{1}{4}$, $\frac{1}{3}$, similar values hold in the case of the intervals less than the octave from just below the octave, through major and minor seventh to major sixth. The intervals on either side of the fifth must, of course, have high values, sinking in the tritone perhaps to the value of the minor seventh, and in the minor sixth to a somewhat lesser quantity, yet still greater than that of the major sixth. But these regions cannot yield the fierce 'collisions' that we get near the octave and the prime. The high value of the augmented fifth lies near the minor sixth as the quick fall of the curve in some way indicates. A zero value may be assumed to lie somewhere in the paraphonic zone[1]. The very small symphonic zones cannot properly be joined to the rest of the curve by lines; for there is no gradual transition from them to their neighbouring diaphonies through an intermediate paraphony. So it would be useless to indicate their relative degree of symphony by vertical position in the figure. There is a transition only where the lines are drawn out. As already said, the ranking that should be given to the fourth is doubtful. The figure, of course, takes no note of any differences of smoothness that may be due to coincidence of partials or of difference-tones, as these are only subsidiary differences that would be ineffective apart from the differences inherent in the tones themselves that the figure attempts to indicate. But the beating of upper partials and of difference-tones would aggravate the original diaphony of the primaries most of all in the immediate neighbourhood of the great consonances, as has already been indicated[2].

---

[1] According to Krueger neutral intervals are to be sought "in a middle region of vibratory ratios close beyond the sevens intervals," i.e. in the first place in such ratios as $6:7$, $4:9$, $5:9$, $7:9$ (31, 252).

[2] Our figure agrees very well with the similar figures that may be constructed from the experimental ranking of the intervals of the octave given by C. F. Malmberg (36, 112) for the 'factors' or aspects of intervals named by him Smoothness (relative freedom from beats), Purity (resultant analogous to pure tone) and Blending (a seeming to belong together, to agree. The curves for these three are in their general course very similar. Malmberg's fourth factor of Fusion (a tendency to merge into a single tone, unanalysable)

Thereby the extent of the zones of original symphonies has perhaps been somewhat reduced; and the point of optimal symphony has been greatly sharpened and emphasised by the smooth system of symphony throughout difference-tones, primaries, and partials that then accrues. In the case of intervals beyond the octave, and, still more so, beyond the double-octave, the ordinary harmonic partials present in instrumental tones would carry forward the basis of original paraphony which we have established for the first octave and the greater part of the second. And the beating of these components would similarly mark out the neighbourhood of these distant symphonies, and so on. In considering the need to appeal at some time to the memory of volumic proportions and their systems in music it must not be forgotten that for the consonance of successive tones no help can rightly be sought either from partials or from difference-tones. The relations upon which we have founded our constructive theory are the only ones that then remain over. Here the step from them to the work of memory is therefore direct. And the memory of volumic proportions is, as we know from vision and from the sense of time, a well established and well trained faculty.

In connexion with successive tones we have already shown how the two great consonances would make themselves felt even in music that hardly knew such a thing as simultaneity of tones. The basis of these successive consonances is exactly the same as that of the simultaneous ones. And to the fifth the fourth is bound almost hand and foot, as a distinctive grouping of tones. For the other intervals less than the fifth—the specially emmelic ones—no clear definition is given in mere succession. Their determination will therefore be decided by two chief conditions : the melodic cogency of the step, which is the greater the smaller the step (no doubt within a certain limit); and a basis of ease of distinguishability between the tones of the step which, we may suppose, is closely akin to the distinctiveness of pitches which produces paraphony

---

shows itself graphically as a pure distance-judgment, correlated to separateness of pitches. Its course of steady decrease from minor second (maximum) to major seventh (minimum) is broken only by a slight rise in the fifth and a rise to greatest maximum in the octave. This indicates again the unitariness of these intervals. Probably Blending and Purity are expressions for the same thing in Malmberg's tables. He also found that, "when by the use of two sets of tuning-forks, the just intonation was compared with the tempered intonation, no difference of ranking of the intervals large enough to affect the order resulted from the difference in temperament" (p. 107).

For evidence of the way in which partials, etc., mark out points in the paraphonic zones where pure tones give little if any differences, see 28, 491 t.

as against diaphony in simultaneous tones. It is in agreement with this that we find in monophonic or homophonic music so much freedom and variation in the subdivision of the two 'tetrachords' left by the fifth and fourth. But as soon as polyphony has revealed its great preference for such thirds as we have adopted as fundamental, all the melodic or successional aspects of music would necessarily come under control. And we have already shown that a basis exists for the (attitude of) apprehension of simultaneity in succession that we find in our music and that is able to make a succession of tones rather a harmonic whole than a melody, or to give to a melodic succession the structural form and the atmosphere of a harmony. So strong in fact is this colouring of harmony in many of our melodies that various theorists have sought for a purely harmonic foundation for melodic figuration,—a procedure which is surely misleading. This harmonic tinge in melodies only appears when the structure of the sequence is such as to invite that sort of reminiscence from a mind already full of harmonic experience. But it never appears even from such a mind when the melody proceeds mainly by small steps. These cogent movements are strong enough to repel any incoherent associations towards dissonance that we might for once try to weave into them.

CHAPTER XXII

SYNOPSIS OR OUTLINES OF INSTRUCTION

THE following may be taken as a summary of what has been expounded in the previous pages. It will also serve as a sort of introduction to the positive results I have reached. There are many readers who will be anxious to see what it all amounts to when the criticisms, evidences and authorities have been laid aside. Besides, no time should be lost in bringing to the succour of the student of music (and perhaps of harmony in particular) what may be firm ground for his intellectual efforts with that art. Until now he has been fed with unexplained empiry or with obviously unreasonable messes of harmonics, both of them diets of no sustenance for the hungry intellect. The teacher of musical theory will perhaps be able to follow this brief sketch of its foundations and to give his pupils a reasonable and continuous introduction to the art. For that reason I have put the sketch into the form of an outline of instructions, in which the topics are merely arranged and briefly stated in their systematic order without being expanded into detail. That extension the teacher will readily make for himself according to his inclinations.

1. The difference between tones and noises is fundamental. Tones are smooth, regular, balanced, symmetrical; noises are the contrary.

2. Ordinary musical tones are really blends of tones (fundamental and other partials or harmonics). Analysis reduces them, by methods that are already generally familiar, to a single series of *pure* tones. Chap. I.

3. Tones differ in pitch, whereby they fall into an obvious order. They differ also in volume : those at the one end of the series are large, massive, voluminous; those at the other end are small and point like. Between the two ends there is gradation of volume. Tonal volume we feel to be a quantitative difference in tones.

4. Volumes commonly consist of parts. The question thus presents itself : does a tone consist of parts? Have different tones any parts in common?

Consider simultaneous tones first. Play c with d, e, f, g, a, b, c^1, d^1, e^1, ... (cd, ce, cf, etc.). Notice :

(a) that the two tones never seem to get entirely away from one another; c and d seem to confuse each other very much, they seem to be mixed up with one another in confusion; c and e seem to be less 'in a heap'; and so on progressively; c and d^1 are relatively free from one another.

(b) that, nevertheless, the series of pitches of d, e, f, g, etc. is the same in combination with c as it is when these tones are played singly in a series; the series stands untouched in spite of the fact that the tones of each pair never become quite free from one another.

5. Now write down the series of pitches as dots (or as points separated by a short space, representing the fact that between c and d, for example, other tones could be distinguished—c♯, d♭, quarter-tones

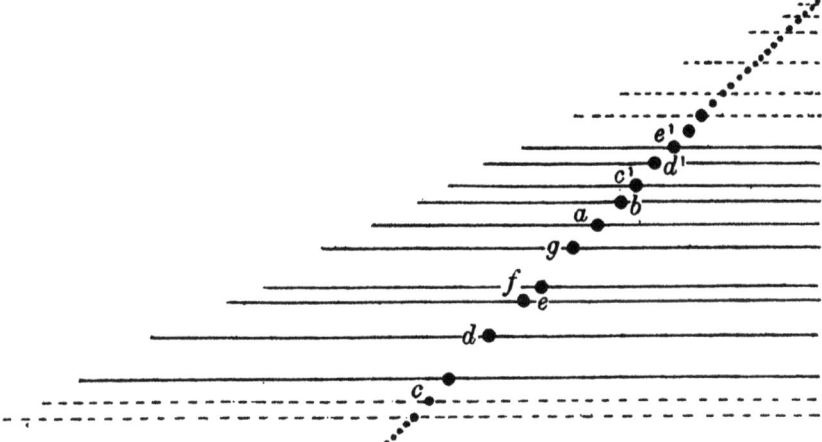

Fig. 7.   Showing the relation between the pitch and the volume of the tonal series.

In this diagram the volume (length of line) of every higher tone must be supposed to be projected upon the volume (line) of the lower tone. They have been separated into an ascending series only for clearness of exposition. The dots represent the pitches of tones.

or eighth-tones, etc.). Then put volumes round the dots to satisfy paragraph 1 (tones are balanced, symmetrical) and paragraph 4a (no two tone volumes are entirely clear of one another). It will be found that the volume of every tone will have to go quite up to the right-hand end of the pitch series where the dots for the highest tones lie. Thus (cf. Chap. II) : Figure 7.

6. Repeat this series cd, ce, cf, etc., noting this time, instead of the gradual change throughout the series, the features that distinguish cd from ce, from cg, from ce^1, etc. For this purpose it may be necessary

to compare the pairs of tones, not as they stand in a series, but in some other order, pair against pair. In cc^1 and, to a noticeably less extent, in cg we observe a balance or symmetry of the whole sound that reminds us of the appearance of a single tone.

How can we represent this with the lines of figure 7? If the line for the higher tone of the pair grows progressively smaller, as is there indicated, we shall at a certain point get this figure (fig. 8) where the higher tone's volume just fills the upper half of the lower tone's volume. This arrangement would surely give the nearest possible approximation to the simplicity and balance that characterise very smooth (or pure) single tones. Hence we may ascribe it to the octave. The *volumes* of tones an octave apart would, therefore, be to one another in the proportion of 2 : 1.

Another form of balance would be given when the two new points of the upper tone lie equally far on either side of the middle (pitch) point of the lower tone, thus balancing each other around it. But the whole would have more points to distinguish it from the single (pure)

Fig. 8.   Showing the relations between the volumes (and pitch-points) of a tone and its octave.

tone than the octave has. Hence it will not be so unitary in effect as the octave. This arrangement (which figure 7 shows must arrive at a certain place in the scheme) we may ascribe to the fifth. And the volumes would be in the proportion of 3 to 2 to one another.

7. Hence we now know precisely how simultaneous tones overlap, and we can correct (or annotate, as the case may be according to the instructor's procedure) figure 7 to that effect. The volume of a tone decreases by half for every octave, by 3 : 2 for every fifth. (These proportions have actually been embodied in figure 7.) These proportions happen to be the same as those that hold for octave and fifth between the rates of the physical vibrations required for these intervals. For all the other intervals the ratios of their volumes follow as a matter of course (just as do the ratios of the rates of physical vibration). It is the balance of volumes that characterises the octave and the fifth as heard intervals in respect of their fusion; they form a sound that is more like one tone than other intervals do : the octave most and the fifth next. The other intervals do not differ so markedly from one another in their appearance of balance as do the octave and fifth from

one another and from all the rest. But they have been graded by careful experiment in the following way : fourth, major 3rd, minor 3rd, major 6th, minor 6th, tritone, major 2nd, minor 7th, minor 2nd, and major 7th. The volumes corresponding to these intervals seem likewise to grow less and less balanced. Their characteristic points come more and more into conflict with one another. But for a fuller account of these grades we must await certain lines of information to be given below (18). Chap. III.

8. The same relations can be established between successive tones (in the matter of the relative positions of their volumes) as hold for simultaneous tones. Only it is not now primarily a question of balance or rivalry. Sequences of tones are not of themselves dissonances (or consonances). But they can play the part of dissonances (or consonances, as the case may be), in music (as arpeggio chords); and they then call for the same treatment. This may be easily explained by supposing that upon occasion we can take a point of view from which we notice the balance of parts formed by two successive tones. These are apprehended in projection against one another as parts of a whole. In fact there is probably a special charm and value in these relations of balance or unbalance, because they are not accompanied by the actual fusion or clashing peculiar to simultaneous tones. Chap. v.

9. Interval is not the same thing as fusion (or as consonance and dissonance). Thirds and sixths are similar in fusion, but very different as intervals. Interval is the constant proportion between the sizes of the volumes of any two tones of absolute pitch. The same proportions can be got in an indefinite number of lower and of higher tones. Then the two volumes are proportionately larger or smaller. Chap. VI.

10. The musical range of pitch (approximately A_2 to c^5, or seven octaves and a third) is the range of difference in physical vibratory rate within which, in the volumes of the tones evoked, there is a constant volumic proportion (or interval) for any given ratio of vibrations. At the upper limit of the musical range the proportion begins to be too large (e.g. $2 : 1 +$ instead of $2 : 1$ exactly for the octave ratio of vibrations), so that the highest tones appear to be somewhat flat; at the lower end the proportion is too small ($1 : 2 +$) so that the lowest tones appear not to be low enough, or as low as they should be according to the ratios of their rates of physical vibration. (These restrictions seem clearly to be due to the limits of size and of subdivision of the responsive tissues of the cochlea). Chap. VII.

11. The predominance of the central point of the volume of a pure

tone and the smooth grading of intensities round that pitch-point set us a pattern for the apprehension of tonal masses in general. Except in so far as melodic interests draw us away, we tend to apprehend the mass centrally, i.e. it seems to have the pitch of its lowest component. When the melodic interests of different 'parts' or voices are equally developed, the bass will be the most effective. For this reason, also, we reckon intervals always upwards; intervals are more readily learnt and recognised in ascending form; descending intervals have to be learnt to a greater or less extent as something fresh; and there is a similar disconnexion between successive and simultaneous intervals. Chap. VIII.

12. The harmonics or upper partials that constitute the blend of musical tones are heard synthetically: i.e. we do not strive to concentrate the attention upon them separately to the exclusion of the rest of the blend; and they are intentionally kept so weak that this will be difficult to do. We hear them then merely as a pleasing change in the 'surface' of a tone's volume, departures from the ideal smoothness and graded balance of pure tone that add interest to the whole while leaving its balance and smoothness approximately intact.

Difference-tones (of which the chief are: $h-l$, and $2l-h$) occur only when two primaries are sounded together. Their presence is generally overlooked, not only because they are heard synthetically in the whole tone-mass, giving it a touch of lowness, but because they generally arise only within the ear and so have nothing in the instrument corresponding to them, and because they appear only when two (primary) tones are sounded together. The separation of these primaries analytically is a task that exceeds the power of (perhaps) most people, so that it is little wonder that the difference-tones pass unobserved. But the chief difference-tones are not hard to hear.

The first and prevailing attitude adopted by a listener towards the primary tones of a chord is likewise that of synthesis, which is founded upon the inevitable overlapping of the volumes of tones (described above). The ease of analysis that is possible for specially gifted ears cannot, of course, annul this synthesis, although it makes the ear aware of the component tones. Practice will tend towards the same analytic ease, which naturally directly subserves the purposes of accurate and beautiful musical production, etc.

Difficult directions of synthesis are favoured by different musical methods, especially by melody ('horizontal' direction) and by harmony ('perpendicular' direction). Polyphonic and harmonic music differs in the prevalence in each of one of these methods as a basis of structure : melody in polyphony, fusion in harmony. Chap. ix.

13. The connexions established between certain chords, whereby they come to be recognised as inversions of a certain chord, are founded upon the (volumic) pattern formed by the volumes of the several over-lapping tones (fig. 9). This pattern is not obscured or distorted by being continued upwards or downwards into the neighbouring octaves. And the continuous pattern thus obtained contains all the different inversions of the 'one' chord.

The octave is the only interval that makes possible such an extension of pattern without any distortion of it. The fifth is quite unsuitable for this purpose, high consonance though it is. This peculiarity of the octave is referable to the special way in which its higher tone fits into the upper half of its lower one.

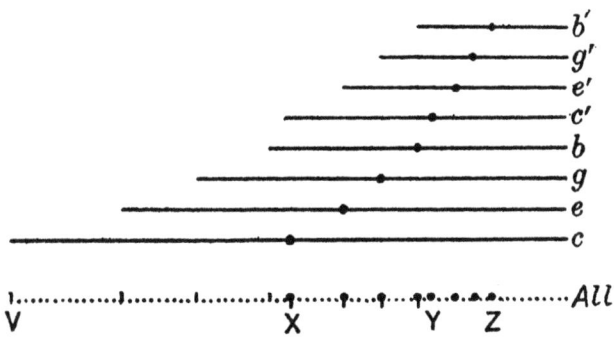

Fig. 9.   The proportions of parts in the 'All' line are repeated from VX to XY to YZ in reduced form (1 to ½ to ¼).

But the connexion thus established between inversions does not in any way annul their very important differences; it only gives them a close connexion in spite of these differences. And this connexion is not a merely formal one, based upon an algebra of transposition; it must always be real, i.e. the listener must feel or recognise the connexion of the tonal patterns with one another or with this pattern that includes them all. Chap. x.

[The instructor may find it helpful at this point to offer the following information :

The organ of hearing is the cochlea, in which a long thin membrane occurs, called the basilar membrane. This membrane stands in connexion with a parallel strip of sensory nerves. The cochlea is coiled up like a snail's shell (without the turn at the apex of it), fig. 10.

Every note that reaches the ear comes first upon the point *a* of the membrane, and affects a length of it that is proportional to the length of the air-wave for that note, or inversely proportional to the number of air vibrations per second. Thus a note of 100 vibrations will affect twice as long a strip of the membrane as one of 200 vibrations, but each begins to work at the point *a*. The point of the membrane most intensely affected lies at the middle of the length affected by any note, and from that point to the ends of the part affected the intensity probably decreases evenly.

Hence flow all the features of tones enumerated in the preceding paragraphs. An attempt may be made to deduce them by way of recapitulation.

Fig. 10. The outline of the sensitive strip in a human cochlea.

The student may find such statements as thus refer to a physical membrane, etc., not only acceptable and convincing, but also preferable to a discussion of what we merely hear :—in spite of the facts (1) that (with proper guidance) tones are immediately observable by him in all their aspects, (2) that he has possibly never even heard of the cochlea before, and probably knows nothing of it by direct observation, and (3) that, being perhaps a music student, he is eager to enjoy or even to create works whose entire essence resides in sounds as we merely hear them.

People of this age are credulous with regard to statements of a scientific form regarding material things, while they will hardly even believe that what they can plainly hear or see or in general have before or in their minds can be properly observed or made the subject of scientific study and proof. When invited to note a description of mere experiences (e.g. sounds and tones) and to follow a train of reasoning about them, they feel that they are being deluded or beguiled into mystical fancies, if the conclusions attained embody results that surprise them. But there is no reason why we should be unable to observe aspects of experiences (e.g. of tones) and by reasoning to infer things about them that we had not previously observed or known.

Even writers on the science of music seem not uncommonly to be almost afraid to trust the verdict of the ear, as if the system of these verdicts that accumulate in the course of time would likely show itself to be erratic and confused, uncontrolled by any sort of law and order.]

14. Melody is constituted not by a mere succession of tones, but by a motional phenomenon or connexion between tones which supervenes upon their sequence, provided their pitch difference and their distance in time is not too great. This motional connexion is somewhat the same as the connexion that turns the successive stationary pictures of the cinema into a continuous picture in which the parts move. Many useful analogies can be drawn between these melodic and cinematic motions. But they must not be exaggerated. Melody is not a slurring or gliding tone; it is a much finer kind of motional connexion than that, and it supervenes even when the sequent tones do not glide in pitch at all.

The motional connexion between tones is the more obvious the nearer the tones lie to one another. The consonance of the interval included in a leap probably improves the connexion. So does any other influence that strongly suggests a next tone, especially a near one. Hence arises, for example, the powerful tendency of the 'leading' tone to the tonic.

15. As soon as music passes the homophonic stage of a single melody with a rudimentary accompaniment, the chief problem is how to make two or more melodies move clearly alongside one another without their arresting or confusing one another's motions. This is the problem of polyphony, and music in two or more voices is well termed polyphonic. It might also be called polymelodic. A clearer term is afforded by the word 'paraphony,' which associates the notion with the familiar word 'polyphony,' but indicates specially that the voices have to move alongside (para) one another, without any mutual disturbance or confusion (p. 155 ff.).

Harmonic does not differ from polyphonic music, as the latter term might suggest, in not satisfying the needs of paraphony and so in not being a combination of voices, but in being rather a new way of putting tones together. In so far as paraphony is concerned there is practically no difference between the two styles. The difference lies in the aspect of tone combinations that is made the ground of the artistic synthesis. In polyphony the melodic outlines or figures of each voice are the elements of artistic construction; each melody must be either at least tense, cogent, and expressive, or at most skilfully thematised, commonly in close relation to the accompanying melodies. In harmonic music this interest has been largely neglected in favour of a great concern for the tonal patterns created at their moments of simultaneity. These patterns (or textures or colours—all these terms suggest a modification of the 'surface' of tonal masses), are now made the chief ground or object of artistic workmanship. Only one voice—commonly the highest —is thematised, the others hardly at all. At times none may be so; then the music wafts along in breezes of harmonic surface or colour. But the needs of paraphony must always be satisfied. And so we must recognise that all music is primarily and essentially paraphonic and only in the second place either polyphonic (in the historical sense) or harmonic (or ultra-harmonic). Chap. XIX.

16. As far as circumstances permit, simultaneous melodies form themselves, or present themselves formed to our attention, just as do single melodies. We do not need to attend to them and thereby to combine their tones into a sequence by an inner act of mind. In fact we do not know which tones belong to the one or other melody until we find them incorporated in it. The melodies flow beside one another like the streams of raindrops that run down a window pane in devious neighbourly courses. The essential and inevitable overlapping of tonal volumes does not seem to create any general disturbance of paraphony. Hence we may infer that melodic motion holds rather between the pitch centres of these volumes than between the volumes as a whole.

This primary freedom of paraphony is characteristic of the intervals of thirds and sixths. Chap. XIII.

17. The other intervals offer less freedom to paraphonic flow. In relation to the series of fusions (given in paragraph 7 above) we find towards the consonant end an increasing effect of (symphonic) obscurity and arrest; towards the dissonant end an increase of (diaphonic) confusion and restlessness. 'Symphony' expresses the apparent unitariness of the interval's mass of sound, its approximation to the balance and unity of a single tone. Compare the Greek definition of 'symphony.' 'Diaphony' indicates the apparent inter-penetration of the tones, by which they obscure one another's pitches and outlines, and so suggest a movement to other neighbouring tones that will give either greater balance or easier paraphony. Symphony (and unison, i.e. a single tone as the momentary identity of two voices) therefore creates a break of melodic flow and a feeling of arrest, diaphony tends to obscure the melodic lines and suggests transition to more peaceful terms. Chap. XVIII.

18. We have already seen how the symphonies of the octave and fifth are created by the peculiar way the volumes of these tones fit into one another so as to appear like one balanced whole or like one tone. In diaphony there is a collision or rivalry of the defining parts of the volumes of tones due to their proximity to one another (thus

all the diaphonies neighbouring on the octave and on the prime or unison) or to their failure by a little to yield the balance of the fifth (thus the small zone of diaphony on either side of the fifth). In paraphony the defining points of the two volumes stand well clear of one another so that the two lines are easily apprehended as distinct even though simultaneous. These relations can be traced more or less distinctly throughout the greater part of the second octave, especially up to the twelfth. The irregular position of the fourth in this scheme corresponds with the somewhat unstable character of that interval in music. Chap. XXI.

19. The effects of symphony and diaphony are, of course, worse when two intervals of the same species follow one another than in a single interval of that kind. Hence sequences or 'consecutives' are forbidden,—the more stringently the more symphonic or diaphonic the interval concerned originally is : octaves (or unisons) most, fifths less, fourths only a little; minor seconds and major sevenths more than major seconds and minor sevenths; or minor seventh and tritone perhaps less than other dissonances. Chap. XII.

But single intervals are forbidden too, or are at least inadvisable, unless they are approached in a way that counteracts or obscures their own symphonic or diaphonic effect. Chap. XV.

20. Of these counteracting or obscuring methods the most important is the use of contrary motion in approaching symphonies and diaphonies. This has the effect of annulling the unitariness and arrest peculiar to symphony and the obscurity and confusion of diaphony. The lines of motional connexion, proceeding in contrary directions either inwards or outwards from the pitches of one interval to those of the next, stand clear of one another and are therefore paraphonic.

Oblique motion has a kindred function. The stationary voice involves only a renewal of the sound already heard. The other voice has then almost the same freedom of action as a single melody. There is no arrest or confusion between the two voices—the stationary and the moving one,—because the one voice moves so clearly and distinctly.

Similar motion probably leaves the intervals quite unaffected, so that their own naturally inherent characteristics of unity and balance (symphony) or confusion and rivalry (diaphony) emerge unmodified. The reason why similar motion is so often forbidden in musical construction is not that such motion is a bad method in itself, but because it is ineffective to remove the undesirable characteristics of the intervals concerned. It is the interval that is bad, not the similar motion.

Contrary motion is specially good because it annuls the confusing effects of symphony and diaphony.

Melodies should not cross or overlap because the motional connexion that constitutes melody tends to establish itself through the nearest or easiest steps. Thus the individuality of melodies tends to be broken, unless, of course, a special character is given to each by its tone blend (or timbre) or by other means such as the special figures or themes contained in the melodies, and so on. Chap. xx.

21. Of the tones of an interval the lower is on the whole the predominant one. If other tones are inserted between the first two, they are less prominent than either of the others. Thus we can readily establish a grading of prominence or exposure amongst the pairs of voices of a polyphonic work. This grading is : B–S, B–A and B–T, S–A and S–T, A–T, where the letters stand for Bass, Tenor, etc. (p. 102 f.) Whatever appears in any pair of voices is the more effective in proportion to the exposure given by the voices. If the effect is bad, it will be worse than it otherwise would be; if it is good, it will be better. Hence arises the gradation of prohibitions according to the pair of voices to which they refer. Chaps. xii, xv.

The prominence of the inner parts will naturally be less the greater their total number. Hence we find a gradual relaxation of prohibitions with the increase in the number of parts. Even the outer parts are less prominent when there are many altogether than when there are few, so that the relaxation of prohibitions is quite general in its degree. Each melodic rivulet is less clear and distinct when there are many of them. The common preference for four-part music probably indicates that in four parts a fair balance is attained between (a) the disadvantage of the unvaried exposure of all details and the thinness of interest in two or three voices, and (b) the disadvantage of the loss of distinction of details and the overfeeding of the interest by more voices than four. Chap. xx.

22. Contrary motion is not the only means of reducing symphony and diaphony to paraphony. Any means is effective in so far as it strengthens the melodic connexions leading through the unfavourable interval.

An obvious device consists in making the motional tension of each part specially strong. That tension, as we saw, is the greater the nearer the tones lie to one another in pitch. Hence at critical moments when an unparaphonic interval has to be passed without the help of contrary or oblique motion, we may help—it is the artist's problem whether that help is at any time sufficient—the motion by step rather than by leap in one voice or in both. As the highest voice is in harmonic music usually the most distinctively melodic (or thematic), the step will be more effective in it. Here again we can easily set up a gradation of cases : worst—leaps in both voices, step in lowest and leap in highest, step in highest and leap in lowest, steps in both voices—best. If one of the voices be an inner voice, the last degree of this series would probably be an unnecessary embellishment (p. 128 f.).

23. The two preceding paragraphs (21, 22) indicate that neither paraphony nor polyphony is to be supposed to be limited to two voices or to any definite number of them. The problems of paraphony run continuously into any number of parts. The effects are complicated only by the greater number of intervals that then appear and the difficulty of regulating them all together to the desired effect,—either melodic motion for a specified length of time or arrest of it at a certain time or a variable degree of this arrestive effect. The difficulty of regulating many parts at once is due to the different symphonies and diaphonies that inevitably appear when a column of paraphonic intervals is formed, e.g. ceg or $ce\flat g$, etc. (We suppose here logically that polyphony sets out from an attempt to combine several paraphonic intervals together. Really the general course of progress went probably rather from the greater consonances, which first drew the greater attention to themselves and were used to embellish a single melody during its course and without themselves forming another melody. The slowness of early progress was probably largely due to the obstruction natural to this starting point. The consonances are the easiest intervals to find but the more difficult to use artistically.)

The resultant effect of such combination depends upon factors already enumerated, especially the nature of the interval formed and the exposure given to it by the voices it stands in. Thus we can arrange the various chords that involve the same set of pitch-names according to the greater stability of effect. For example, of chords involving the tones c, e, and g, ceg is the stablest and most arrestive, because it contains a fifth (consonance of second grade) in the outer voices and the most fused paraphony (major third) in the outer voices. Of the two inversions the second gc^1e^1 is better than the first egc^1, because the former has major paraphonies for the latter's minor ones and the fourth is less noticeable in the upper voices. Similarly in chords containing the tones b, d and f the worst or most confusing arrangement is bd^1f^1, then fbd^1, while dfb is the best.

24. Because of the general trend of these differences and also because of the theoretical interest of columns of thirds (major or minor), the arrangement of the tones of a chord that gives a column of thirds has been called its root-position. We can readily think into that term the notion of greater stability, arrestiveness, or paraphony. But it is wrong to suppose that any chord has been in any sense generated from any one tone, as the older theories that operated exclusively with harmonics supposed. But a small chord, e.g. bd^1f^1 may well come to be felt to be a part of another larger chord (gbd^1f^1) in so far as the latter (or its 'pattern') includes the former and comes to be so important for music generally as to attach the former to itself as a mere part or reminiscence. But this connexion, if it is to be valid, must be really felt.

When symphonic intervals are combined with diaphonic, we find that the loss of distinction due to the great unitariness of the former is largely annulled by the loss of distinction due to the confusion introduced by the latter. Consequently single and second symphonies (octaves or fifths) may pass unguarded (without the help of contrary motion, etc.) in discords.

25. A special paraphonic difficulty is created by the interval of the fourth, especially when it is exposed by resting upon the centre of a sound mass—the bass. The fourth is not a dissonance, as has often been asserted. It is what we already know it to be—the lowest grade of symphony. But especially when it is exposed rhythmically as well as phonically, it seems to call for or to suggest the neighbouring major third, and so to claim a resolution, as do the dissonances. If this suggestion is to be avoided, exposure of the interval must be reduced as far as possible, and the melodic line leading through the chord in which the fourth appears must be made as cogent as possible, especially the line through the bass. If the suggestion towards the third is welcome, it is increased by rhythmical exposure and so a special progression—a (symphonic) cadence (or general arrest of melodic flow)—is attained. Chaps. XVI, XX.

The case of the bass-fourth is probably only one of a class of similar cases. Another example seems to be given by the minor sixth, which has been held to be dissonance, although it is now generally treated as a consonance. But the interval identical with it in equal temperament —the augmented fifth—is always a dissonance, probably because the augmented triad so strongly suggests the major triad that the former appears as a distortion of the latter.

This reason for the difficulty of the fourth from the bass explains why the effect of such a fourth is not obliterated in a discord, while a second fourth in the same voices (consecutive fourths) is relieved, if either the second (or even the first) stands in a discord. In the latter case the fourth acts merely as the symphony it is; in the former its effect is due to the call for the major third excited by its exposure on the bass.

26. The tendencies and possibilities of resolution of dissonances are not so much the result of any sort of connexion between different intervals, as far more probably a general expression of the most cogent melodic movements that will satisfy the propulsion of melodic move-ment that is characteristic of diaphony. The most cogent movement is in the first place the shortest (in difference of pitch), but due regard must be paid to the limitations imposed by key structure, the paraphonic nature of the resulting (following) interval or of the whole chord in which it stands, thematic form, change of key, etc. Apart from these determining conditions there are probably no limits to the possibilities of resolution.

27. It is not to be supposed that chords are essentially different from intervals. Both of them are—in music—primarily paraphonic structures, which differ only in the number of voices involved. The putting together of three voices is only an extension of the task of combining two melodies, not a new kind of task. But we learn to look upon the structures formed by three or more voices as typical volumic patterns, which we reduce to a few comprehensive types. The relation of inversion, for example, forms a very important step in this reduction. So accustomed do we become to this procedure that we largely cease to be able to look upon intervals (of two tones) except as fragments of chords. Nevertheless there was a time (for music generally and for each musical person) when intervals had not yet come into these relations. The final result of long experience with the most frequent types of chords is the formation of a system of chords and their relations and progressions, commonly known as Harmony.

The artistic concentration of attention upon harmonic rather than upon melodic form and relations leads to a special systematisation of chords by means of which they are grouped round a centre (tonality) with parallel sub-centres (keys). Tonality reacts upon the scale, moulding it so as to make it yield the biggest and most coherent system of chords possible. From this diatonic harmony the system is extended gradually to a system of chromatic harmony and to other types that are at present insufficiently defined.

28. The development of tonality raises into importance the chords that are related to one another in certain special ways, especially those that stand at the interval of a fifth or fourth from one another. Here we have in a new form the relation of balance characteristic of the symphonies; the octave virtually drops out because the repetition of a chord an octave away is—in all but its 'brightness' or volumic size—the same chord: or the octave is just the *greatest* balance, if you like. The fifth (or fourth) is the next, and therefore the main, pivot of all tonality. These points of tonality are also the only ones that repeat a chord exactly without any distortion. (Possibly this balance of chordal relations at the fifths is responsible for the final form of the scale.) So we come to think these points easily together. And their connexions can thus combine with the circumstances already indicated to counteract the effects of symphonic and diaphonic intervals towards any special end.

There is some evidence that this power of harmonic connexion varies with its closeness, being greatest for the fifth-fourth, next for the third-sixth, and least for the second-seventh (or no) relation. But the matter probably requires further justification before it can be taken in this general form. For the first grade of harmonic connexion there is no doubt of the existence of a modifying power.

29. It must not be thought that the laws of primitive music are annulled by its development. They are rather merely complicated and fulfilled. Rules of harmony that bind one age are not disregarded in the next. Nor do we merely learn to tolerate what a previous age forbade. We rather learn to annul or to transmute the effects that once created strain, by bringing to bear upon them influences that the previous age had not discovered, but that are essentially the natural products of musical development, or, better, of tonal artistry. We bring the once jarring phase into a setting in which it forms a proper extension of those very laws that were previously the just ground and reason of its prohibition. Music is an orderly realm of beauty, intelligible in every aspect; it is not a temporary code of arbitrary preferences that extol themselves as art.

The main thesis propounded in this book, so far as the actual structure of music itself is concerned, is embodied in the paragraphs of the preceding summary from no. 14 onwards. It may be briefly stated here :

The fount and origin of probably all music whatever is melodic movement, or simply, melody (not limited, however, to thematised melody). Primitive music is monomelodic, embellished by reduplication at some interval or with irregular heterophonic accompaniment. Modern music, whether it be classed as polyphonic or harmonic, belongs to a great era of polyphony, of which the essential problem is the construction of concurrent melodic streams that will leave each other's motions unimpaired and produce effects of arrest as they may be desired. The solution of this problem embraces the solution of the nature of similar, oblique, and contrary motion, of the prohibition of consecutive intervals of the same kind as well as of 'hidden' consecutives, of the puzzle of the fourth from the bass, of chordal structure in general and of the differences from one another of a chord and its inversions, of the need for resolution of discords and its tendencies and possibilities, etc.

Or : it is commonly held that our music is "the concord of sweet sounds," a structure of which the fundamental material and ornament is consonance in its special or more perfect grades. We might rather say that music is a 'concourse' of sweet sounds, in the literal sense of the word, a structure in which movement or melody is the fundamental material. There is rather an opposition than a kinship between the functions of melody and of consonance. The latter is the principle of arrest, inimical to the free course of melody. Music has been created rather in spite of consonance than by its help.

Thus the Greek definition of consonance is upheld both in its connotation and denotation. Symphony reveals a loss of distinction of tones in apparent unity. The neutral intervals of thirds and sixths must be recognised as paraphonic; their tones are freely distinguishable. In dissonance or diaphony we find a loss of distinction in confusion or harsh collision. Thus we give new significance and confirmation to the classification of intervals proposed by Gaudentius. Music progresses by greater differentiation between intervals and by the discovery of further factors that modify their paraphonic properties, not by a gradual lowering of the border line of consonance.

CHAPTER XXIII

THE OBJECTIVITY OF BEAUTY

REMARKS have been made at various points throughout the preceding chapters in deprecation of the common tendency to look upon beauty as a matter of convention, or of merely subjective personal fancy, such as prevails in one person or age and may be rejected in another. One feels a certain reluctance towards passing from the study of the foundations of musical beauty in detail to the exposition or defence of a general thesis such as that of the objectivity of beauty. Those who study the detail closely will surely come of themselves upon the general point of view, and will feel its cogency. And there is a great pleasure in thinking out such conclusions and in seeing how they extend their satisfying graces into the whole of one's world. On the other hand there is a certain amount of laborious thought involved in these visions, the communication of which may help to lighten the work of others. In any case, every one of us has his own particular world to fill with strength and joy, so that there is no real risk of robbing any man of the pleasant excitement of that task.

The age we live in is in many ways strongly coloured by the outlook of subjectivism. We have passed through the time in which the first great generalisations of biology were made. Each type of creature has been shown to be the product of innumerable influences that not only work upon it now, encouraging one individual and eliminating the other, but that played in the same way upon the infinite series of its ancestors. The choosings of chance have made us as similar or as different as we actually are. Our members, their forms and functions, our innermost structure and processes, in short every infinitesimal part of us is to a greater or less degree peculiar to us, unique, sifted off from the great drift of life's variations, and stamped 'individual.' How, then, can we expect to be so fundamentally akin that the world should be the same even for two of us? Why, our very eyes, our brain, and the minute chemistry that is its change, have been selected from myriads of possibilities! Even beauty has been held to be merely contributory to life's functions, a subjective accident enlisted in the work of reproduction to be a symbol of the fitness of male and of female to one another. And we make our arts out of these symbols by wandering from the straight path of life's purposes and by perverting a good function to a useless play. So we come to look upon art as perversion and upon the artist as an unpractical trifler, stifling the true functions of his mind—which should serve the great ends of life—by combining them into vain display.

But the artist reaps the inevitable reward of his perversion. He flatters himself with the thought of progress: in the end so many revolutions are wrought in his art that the nothingness and unreality of all of them become too obvious to ignore. Art then comes to be the whim of an age, the latest device to thrill and to dazzle a mind that is 'real' only when it passively acts as a drop in the incoherent drift of life, but that, blowing itself out to some semblance of intrinsic worth and orderliness, shows a brief iridescence of meaningless changes and bursts into dust.

And then the ruin of art begins to infect the other parts of mind. Thought is even degraded to be a mere scheme of symbols, useful for our practical life, but worthless in themselves. We are shut up within the impenetrable wall of a senseless 'mind' without any means of testing the true worth of the bonds that seem to link us to a world beyond, and with every 'reason' to suppose them illusory. Morality becomes the preferences of a social group, and the world-race of nations is then to the strongest, who grants himself all the sanctions of subjectivism. In the vast struggle of these years we are not striving merely to make room for every man's illusions against one tyrant's ambitions. That would really yield the senseless chaos that freedom seems to the enemy to be. We are rather labouring to fulfil the demands for obedience of laws that claim objective worth and that promise to give our souls the dignity of reality, of being members in a coherent intelligible world.

Another source of subjectivism was created by the critical philosophy that has been lauded for a century and a half as the greatest revelation achieved by thought. It is true we must examine the work of knowledge with all care to see what it can validly establish. But if our conclusion be that true knowledge is so overlaid with the devices of the mind itself that it is impossible to say anything about the material upon which it started to build, we are driven therewith upon the rocks of subjectivism, no matter how zealously we repeat to ourselves: "what the mind once claimed to know (really), it may still claim to know (virtually)."

Of course we are glad to have cast away the crude illusions of knowledge that always bind our thoughts so long as we do not examine evidences carefully and reason with caution. In a sense, however, these illusions are only the natural result of a force of knowledge that is eager to work and has not yet learnt to harness itself to the ideals of completeness and security. But if our critical study of knowledge is to lead us to suppose that thought itself is a deceit like a kaleidoscope, revealing nothing beyond its own gyrations but the existence of light itself, then have we cause for misery. We shall be prone to think that the older 'dogmatisms' that knew nothing of critical scepticism, were really sounder products of the spirit than the critical philosophy itself. They may have known all that we know of evidence, consistency, and system, although they were greatly deficient in facts and left in their expositions too much to the imagination and understanding of their readers.

At any rate a sounder vigour is beginning to show itself openly in the sciences of biology and in philosophy. We now see that chance of itself does nothing, and that selection merely selects or favours one of many possibilities, each of which is founded upon the coherence of objective forces that we call laws of nature. All the possibilities of inanimate or animate combination are realised, if only the forces by which they are surrounded tolerate them; and they endure in proportion to their stability and coherence. Every individual, no matter how special or 'rare' he be, is founded upon the not innumerable forces that make an infinite variety of combinations possible. His unique composition involves only the limited gamut of the world's scale. In respect of generality and immutability the underlying laws of life take rank beside even the laws that rule the great bodies of the universe. These are the great chimes that tell the years, those are the fleeting colours of an orchestral hour. But they both go back to the primary balance and beauty of pure tone.

Our study of music has shown us that order and systematic relationship are the ground and basis of the art, not capricious selection and convention. We have indeed to learn to combine the fundamental forces of tone; but there is no caprice or chance in the forms that please. Chance pertains only to the experiments that we shall at any time make, within a certain scope allowed by the age in which we live. We cannot yet adopt what may seem beautiful to a later age. Not that we may not already see some beauty in it; but the orderly scheme of beauty we have thus far gathered does not yet offer it any welcome.

No place has been prepared for it within the art. Music has to grow till it is able to comprehend each special beauty as its own, just as the mind grows gradually towards the wisdom of its years. But both the beauty and the wisdom of a later day are the true fulfilment of their earlier forms, children of the same elemental order and purpose.

It is evident enough that there is a structure and necessity in what we find to be beautiful. That is true in broad outline at least, however difficult it still may be to carry the demonstration down to the finest details. Beauty is of our creating only in the same sense that the forms of life are the result of selection. It comes before us and we find it good. We combine its simpler parts as we can, and we then find laws in the conjunctions that please us. We use for art a material that has already been wrought from the ore, as it were. In short we discover beauty just as we discover truth or even goodness.

Beauty is not beauty merely because we feel pleasure with it. It is true we should not respond to it or feel it but for pleasure. But that is really a tautology. It means only that if we did not respond to it, we should indeed not do so. On the other hand, were beauty not what it is, it would excite no feeling in us. And mere feeling of a false kind never makes beauty. We have to adjust our souls so as to feel aright, just as we have to be careful to think truly. Feeling is merely the thrill that beauty strikes through our being, a symbol of our fitness to share its high functions. We are aware at such times that our experience is most cogent and coherent, as it is likewise in true thought and deed.

Moreover, beauty reaches back continuously into the physical world. The tonal volumes and pitches that are our own experiences, find an objective counterpart or parallel in the excitations spread over the sensitive surface of the ear. And the simplest parts of these at least depend upon equally regular processes in the aerial medium. Although the three stages of the physical, physiological, and psychical are not precisely alike in all respects, yet their differences are completely comprehensible and regular. Except in so far as we cannot yet discern the inherent continuity of mind and matter, there is no mystery in their general continuity of action, as far as hearing at least is concerned. What comes to us as beauty is, therefore, founded upon the order and regularity of the natural world. In fact it is itself a part of the natural world.

If anyone cares to read this conclusion as "beauty is physical," no great objection could be urged, so long as the meaning implied is not altered to suggest that we understand the continuity running through matter, body, and this 'physical' mind. *That* we manifestly do not yet understand. If our comprehension is to be secured by setting down the realistic constructions of ordinary knowledge and of science as mere classifications of phenomena or as mere symbols or counters for thought, we are surely on a false track. No one is really inclined thus to sacrifice the results of science as illusory. Nor has anyone succeeded in basing what we know in science in the broadest sense upon what is exclusively phenomenal to us, upon sensations, for example, and upon nothing else. Nor has anyone yet shown that the entities posited by science are essentially of the nature of the phenomenal units of our sensory experience or the like. What the elements and complex units of the physical world really are in themselves apart from their order and regular changes, we do not know. If we pretend to know by trying to believe they are all illusory or fictitious ('reasonably fictitious,' of course), we only deceive ourselves. For we can never really believe that. The well ordered panorama,—all of which to the last particle we can see with direct gaze,—that some would have us believe in, is a myth. We cannot in ourselves be all-seeing and all-wise. And if we see only in part, and that with great labour of concentration, then there is much of the world that reveals its movements and forms to us only through our own very limited panorama of phenomena. Thus there must be many things in the world whose inner being remains indefinitely hidden from us. Beauty may be a part of the physical world, if you like, continuous in action with it; but it is not a part of that real world which we call physical and whose inner essences are hidden from us. For we have the full being of beauty before us, however difficult it may be for us to understand it completely. We feel it entirely; it is there before us in all its native essence.

If beauty is thus objective, a part of the natural world, we have a greater satisfaction. What we have before us is the promise of all. Beauty then belongs essentially to the world. And the world we may look upon as really so beautiful as we naively observe it to be. Its charms are not the mere dreams of a creature of chance, dreams that will vanish with its death. We may be sure that, wherever there is a mind to view the sunset or to hear a song, provided that mind has grown to the stature of the task, there beauty will be evident. There is not one world for you and another for me, but one—according to our varying powers of apprehension—for all.

In pre-critical days every discovery of awakening science was apt to be hailed as new evidence of design in nature and of the presence of God's hand in the world. We should not generally now be so simple as to imagine that law and order form a sufficient proof of dependence upon an omnipotent person. But for all that we cannot forgo our wonder and admiration at the awful complexity and the marvellous stability of things. No amount of belief in mechanism will render us quite cold to these effects.

We may indeed tell ourselves that our minds, having been evolved in dread of nature's rough forces and of the savagery of beast and man, are only too prone to be moved by what is vast and great and complex. And we may dispassionately infer that our notion of a natural body is as of a thing we might ourselves have tried to make—a machine and its works; and that we pass therefrom merely by association of ideas to a notion of the effort and ingenuity that the works of nature would have involved, had a mind like ours devised them. From this point of view we read the effusions of the earlier philosophers of nature with a certain amount of condescension.

We are supported in this attitude by the reflexion that though mind itself may feel orderly and lawful, such a feeling is but illusion. Really it is merely subjective and accidental, a product of the subtlest cerebral chemistry or physics, not only worthless as a sign of manifest order, but obviously quite beyond any hope or possibility of analysis. We can take as our test of mind's action no formation inherent in itself, no continuity of process that binds it to an objective world, but only the success of its functions, which in its turn is evidenced only by the mind's tolerance of them. Not only can the existence of a universal world not be proved by mind, but we must finally admit that whatever the mind is content to accept is acceptable (or 'true'). There is not only no permanent world from which the world might seem to acquire order by merely reflecting it, but the mind itself cannot be expected to generate order. Its autonomy is sheer heteronomy, its sanctions are its own acts, its only law is mere function.

But if law and order are discovered and proved for mind itself, what then? We may expect to hear the reply : what you have shown to be so orderly is not mind but nature; or, you have merely carried through a useful, helpful scheme, which has no further validity than the brevity and expository value you have shown it to possess. Both of these interpretations will surely in the end prove tiresome and nauseous. For as we all suppose ourselves to have some sort of mind of our own, we shall hardly consent to abandon it all to nature; nor can we enjoy the jest of a pragmatic world for ever. We shall gradually recognise that the order of as much 'mind' as we at any time understand is akin to the great 'design' we now see in nature. We shall become convinced that 'mind' is of a piece with nature. A great and universal objectivity runs through both.

Shall we then be more inclined to see mind in nature,—a vast thought, actually delineated by our sciences, 'thinking' itself (or whatever other faculty of mind we may conceive it to be), forwards by its rotations and syntheses in its own vast sensorium (or conceptorium or effectorium), although we can in no way discern the quality of consciousness that pervades it? Or shall we humbly infer that our own minds are a mere mechanism too, rolling on because it once got started on its futile gyrations, endowed with a stability that is a synonym for 'it is what it is'? Shall we not rather learn to see that, in mind, not only is the essence of order and of law actually made manifest, but inner coherence and stability has even become the ideal of all its effort (pervading the great faculties of sense-beauty, concept-truth, and emotion-joy, love, goodness)? When these ideals seem attainable, its very being is rapture; it is misery so long as they seem indefinitely elusive.

Whatever our general argument or our detail of knowledge may be, we shall probably all incline always to some sort of an idealism, however much of reality we may at the same time find around us whose kinship of being with us we cannot fully prove, at least as regards the last essential phase of that proof. And we may do well to be so inclined. It is not to be expected that a creature of this world should be willing to sacrifice the only direct vision of being it possesses for any of the forms of other beings that its knowledge may describe. Until we learn that mere form can constitute a being, we shall hope in our modesty that the things of this world are essentially (or in their inherent being) greater (spirits) than we are, and in our pride that they have in themselves some portion of the insight of mind and some glow of the emotion we find in ourselves.

The knowledge we have won may at least make us hope and believe that in beauty we are in direct touch with a real aspect of the world. And if the world appears to us in the order of law and in the harmony of beauty, why not also in the kingdom of goodness and love?

CHAPTER XXIV

AESTHETICS AS A PURE SCIENCE

AESTHETICS as a pure science is simply a part of the pure science of psychology. So all the characteristic features of the latter notion appertain to the former.

Psychology is concerned with the parts and the structure of what is commonly called 'mind.' That includes, according to the doctrines of modern science, not only such things as memory, thought, and emotion, not only the bodily feelings of hunger, pain, warmth, posture, and movement, but also all sights and sounds, distances, forms and spaces. Of course some folks object to having these last things called sensations and experiences. But that does not matter at all for the present, so long as we have good reason to group them together and with the bodily feelings just mentioned. All these things call for systematic study together.

In fact, they seem to form a distinct world by themselves and apart from all that science can tell us of the physical forces that arouse pain, warmth, sound, sight, or of physical form and space. The ordinary man lives in his world of sensory feeling as he finds it; he does not concern himself with the physics of sound, light, or warmth, or even of space. Psychology is the name of the science that has to arrange and to explain the parts of this world in relation to one another, however it may afterwards (or otherwise) join hands with any other science (say physiology or physics) to show how feelings depend upon the body or upon matter. This systematic work of psychology has to be done without filling up gaps of ignorance by disquisitions on the relation of the feelings to the body or to matter, and without giving theories about feelings that are merely inferences from what we know of matter or of body (unless the temporary deficiency of our knowledge makes such an effort inevitable). That is what is meant by speaking of *pure* psychology.

The work of pure aesthetics is similar—to give a full description, systematic arrangement, and so an explanation of every work of art in so far as it is directly before the spectator's (or artist's) mind without any regard for all the (otherwise so important) facts and questions relating to the work of art as a physical thing (marble, canvas, paint, etc.) or to its effect upon the eye, the ear, the nerves, the brain.

It has taken science very long to find out that the brain is the organ of mind, so to speak. Most people have still only the vaguest idea of the connexion between the two. But this ignorance in no way alters the efficiency of their minds. They can observe their impressions and the workings of their minds quite clearly; they know whether they feel pleased or displeased, with reason or for some motive, whether they are sure of their conclusions or uncertain, whether their conduct seems to them good, indifferent, or disgraceful enough to need concealment. Of course, they do not therefore know everything about their minds; they may deceive themselves in feeling, knowledge, and conscience; but yet we all know and believe that if they set their minds more vigorously and honestly to work, they might improve its action indefinitely. In the same way the finest art was in existence long before men knew precisely of what marble and paint consist or how musical instruments produce their tones. A composer accepts or rejects a tone for a melody or a chord for a progression, not because he knows something of its physical nature, not even because he knows what partial tones it contains or implies, but merely because it is beautiful and fits beautifully into the work at the point in question. He has all the material directly before his mind by which to judge of this without needing any scientific knowledge at all.

It is the task of a pure psychology in general and of aesthetics in particular to give as connected an account as possible of all that thus goes on before and in the mind.

There are many, however, who say that though much can be found out about the things of the mind, the account can never be made complete or closed. That we shall see in the course of time, if we try. We have at least a great promise of approximate completeness; for do we not expect the work of the mind and of art to grow more and more consistent and perfect? So, to begin with, we have perhaps even more cause to believe in the uniformity and inherent consistency of mind than of nature. Whether mind is in some region or process chaotic, we shall have cause to believe if we fail utterly to comprehend some portion of its ways.

But there are mysteries both in matter and in mind. The darkest mystery of the physical world is why or how it ever was created and set agoing. The very question is a sort of nightmare, through which we help ourselves by the recital of beautiful myths. The problem of the origin of mind is just as baffling. Science evades these ultimate issues. Its only concern is to reduce the world or mind to its simplest foundations. But, given these and what it learns of their nature in the course of analysis, it hopes to show how the complex world and the mind we know result therefrom, without the assumption of any interference 'from without.' We do not now suppose that God steps in, as occasion demands, to wind up the clock of the physical universe or to mend its works. Nor need we suppose, once the elements of mind have been given by the brain it is dependent upon, and the process of complication and interaction of these elements has begun, that the brain here and there exerts its influence anew to keep the process going.

This last idea may be illustrated in a crude way by reference to the problem of consonance. Instead of supposing that there is something about the mass of sound constituted by two tones an octave apart or by the common major chord that we can discover, describe, and set up as the object or cause of the pleasant feeling we have in connexion with it, some folks speak as if we heard the two or three tones precisely as we hear each of them separately (only, of course, now together), as if there were equally little or nothing in either case to account for any unpleasantness, and as if the brain, so to speak, wired a separate message to the mind in the code-word 'pleasant feeling,' which being somehow decoded by the mind would mean : "the tone-group dispatched by separate route is to be 'pleasant,' i.e. runs nicely through my nerve centres."

If this sort of thing were true, the only problem of psychology would be to make a rapid catalogue of the brain's code-words, and to try by all sorts of experiments to catch the brain in the very act of sending off a code-word and to see what was going on in it at the moment (that was nice or nasty). And as a matter of fact that is exactly what the program of psychology involves according to the present notions of (probably) the majority of its professed exponents.

Whatever limits lurk hidden in the notion of pure psychology or aesthetics, that kind of idea must surely be a travesty of mind and art as we know, or, at least, 'feel,' them. It is absurd to suppose that our minds are the mere puppets of a brain-and-matter show. If that is the sort of world we live in, the sooner the silly play is over the better. The opposite extreme is far more worthy of the vast and orderly cosmos we know the world to be, at least the world of nature. Let us suppose that in the mind we shall find the greatest inner order, coherence, self-sufficiency, completeness that is anywhere detectable at all. Let us never rest till we have gone far towards realising this ideal by the most stringent methods of science. And let us then learn to see the great world in the light of the little mind we know, permeated and bound fast by inner forces like those that call us forwards,—truth, beauty, goodness,—although we know the stability of these cosmic aspirations only in their outer forms, not at all in their inherent essence or 'spirit.'

The pure psychology that has been expounded in the preceding chapters is at the same time, in so far as we have confined our attention to the foundations of music, a pure aesthetics. We have found a basis for the art that at no point transgresses the range of what is before the mind of the creative or appreciative musician. No gap in our knowledge of the foundations of music, as we hear and enjoy it, has been filled by our knowledge of the physical processes of sound or of the bodily functions of hearing.

Of course the results attained are in considerable part new; they do not embody at all what the musician may himself have thought of music. But that is no real difficulty. Many an appreciative listener enjoys a new work of music without being able to analyse it or to say of what parts it consists; and yet no one would suggest that the work was not completely before, or accessible to, his mind. In the knowledge of a thing there are two important parts—the knowledge and the thing. And two relations of time are possible between them : they may come into existence practically at the same time, or the thing may precede the knowledge by an indefinite period. This latter alternative holds not only for the objects of nature, but in some cases for the things of the mind as well. It is a familiar fact that various difference-tones have been discovered only comparatively recently; and, before the time of Tartini or thereby, these were quite unknown. Similarly partial tones were not heard and known, but only heard, to be in timbre before Helmholtz's time[1]. And if my exposition is on the right lines, I may claim—to the best of my knowledge and belief—that no one knew

[1] In this sentence special emphasis should be laid upon the word timbre. Gevaert (16, 118 f., 153 f.) is convinced that the existence of the first harmonic was known to Aristotle. Cf. Gevaert's translation (16, 19): "Pourquoi dans la consonance d'octave le grave reproduit-il l'intonation de l'aigu, tandis que l'aigu ne reproduit pas l'intonation du grave?.... Et (en effet) la corde *hypate*, au moyen de sa division (en deux parties égales), produit deux *nètes* distinctes." "Does not Aristotle in his treatise on the Soul say: ἡ φωνὴ συμφωνία τίς ἐστιν (tone is a sort of chord)," (16, 119)?

before what tone and interval in their various aspects really are, although every musical mind must have felt what they are a countless number of times previously.

Our study of the foundations of music has now brought us clearly within the range of the musician's present interests. It is unnecessary for him to know precisely the nature of tone or of interval in their simpler aspects, or to understand the nature of the blend of tone (timbre). But no musician can produce good work without knowing, as well as feeling, the effects of the various intervals upon the flow of his melodies or harmonies. The analysis of these things is the lore of the music student that we have freely used as a basis of induction. By connecting the results of induction with the psychology of tone and interval in their simplest aspects we have made it possible for the musician to follow the growth of music from its foundations in mere tone up to the strands and links that bind tones into coherent music. The study can now be taken over by the musician himself and pursued without the further help of psychology, in so far as this is the general and systematic study of all kinds of sensations and experiences.

No doubt there is much still to be done, the results of which may reflect greatly upon the disposition of the ground thus brought under control. But the musician should no longer feel that he has no firm place upon which to build the systematic structures that seem to be needed for the housing of his musical analyses. We have now a clearer view of what music really is; and we can venture to say that we now understand much of what heretofore seemed to be so mysterious in the art. In the following we shall review a few of these old problems of music that are at the same time special forms of the general problems of aesthetics.

There is a special ease and suggestiveness in claiming for music the need for a pure aesthetics. It has been asserted in a much quoted sentence that "all art tends to the condition of music" (quoted from 33, 227). By this is meant that music is an art that is obviously pure and disinterested, and that the other arts can only with difficulty reach the level of purity that is natural and almost inevitable in music. But all arts are inherently pure; they all are only 'for art's sake.'

Probably the chief element in this distinction is the fact that, while it is very difficult to exclude from the arts of sight the representative functions of vision, there are so few representative processes in tonal sound that the arts of sound remain wholly unrepresentative. How few of the things around us make a tuneful noise! Writers on the theory of program music (e.g. 39; cf. 42) have been able to compile only a brief list : the cuckoo and the cock are alone unmistakeable; the nightingale, the thrush, and some others sing in a way that can be imitated and recognised with fair ease; but the sounds of most things are really noises that are so hard to imitate on musical instruments that their recognition cannot be assured. Besides, these representative sounds offer so little scope for artistic modification and construction that their introduction into a musical work is obviously inappropriate and produces a ludicrous effect.

But there is an aspect of the question that seems to be of much more general importance than these. Sounds cannot really be said to represent the animal or object that emits them, any more than a cloud of dust represents a motor-car, or a trail of steam a railway engine. The form of any object represents it or *is* it in an intimate way that characterises no other effect produced by it. For in all these the form presented in the effects differs entirely from the form of the object itself. In the case of sound this is especially so. The formal aspects of sound (tone, volume, and interval), and the lines of melodies have no relation at all to the shape of the objects that emit them. They cannot then properly represent them. But the pictorial representation of a house or of a man does not merely make us think of the house or the man; it is (pictorially) the house or the man.

It is in no way the merit of music that it excludes representation. It does so merely because few things produce more than a characteristic noise or timbre, and those features of sounds are the hardest of all to reproduce exactly. In short music cannot be representative. Pictorial art, on the contrary, cannot readily exclude the representative aspects of things without sacrificing much that is of the highest interest and so impoverishing itself greatly. It can, of course, limit its scope to the purely ornamental; but valuable and beautiful as that is, our interest in representative forms is so intense that we constantly long to see the forms of pure ornament vivified by the use of representation.

There is, therefore, a comparatively limited field for the art of sound to build upon, great and beautiful as are the edifices it raises thereon. Between this field and the emotional life that is expressed in the art there is a direct and continuous transition. The claim of music to be the standard of all art implies that in the pictorial arts all the ranges of experience other than those of colour and form that enter into artistic appreciation are unlawful intruders. Their presence is, as it were, a concession to our prevailing habits of mind, not a proper part of the art's action. The probability of this being true is, to say the least, extremely small.

For it is the peculiarity of the aesthetic attitude that it is always direct; it appreciates an aesthetic object only in so far as it is before the mind, not at all in so far as it is representative.. Of course, representation is not excluded; but the material thus brought before the mind is of value only in so far as it is what it is, not in so far as its purely representative function is concerned. The primary basis of music is undeniably the tonal material and all its forms and movements. It is not easy to make these representative without effort; but beautiful music might conceivably please merely because it reminded the listener of some beloved person who once made such music or because it indicated much skill and energy in the player. Many concert-goers are inclined to judge musical performances from the point of view of their acrobatic perfection and virtuosity, if not even from that of the loudness of sound attained. In the same way the mere colours and forms of ornament may suggest many extraneous ideas to the spectator. But neither of these deviations from the artistic attitude make us desire to get rid of the suggestive sensory material from the arts. It is not forbidden that a piece of music shall be difficult to perform, but only that it shall not be performed so that its technical accomplishment stands forth as the chief object of interest to the discredit of the inherent beauty of all its sounds. The patchy bedabbled canvasses that are often seen in exhibitions err in a similar way by making the method of work so obvious that the eye is distracted from the intended beauty. But no one would desire that the method of work should always be so hidden away as to be undetectable, merely because there are many who might rather observe it than the picture's beauty. Why then should we wish to banish the representative functions from visual arts because some folks may take the work as an illustration and not as art?

It is surely a great mistake to try to limit the basis of construction of any art to a certain range of 'experience,' even if it be to the sensory range. On what grounds shall we fix our limits? Take, for instance, the space of vision that is so important even in the flat arts of sight. A solid sculpture seems to be the only solid type of artistic object. A picture cannot really be solid; its space is only indicated; if it were not so, we should take it for a solid 'panorama,' as we see a room reflected in a mirror. The picture's space is only perspectival and 'tonal,'—in both forms 'suggested' to the eye, not given. But space is one of the most powerful interests of pictorial art,—atmosphere, distance, roominess, etc. Is it for that reason a foreign element in the art? Certainly not : for the simple reason that the spaces of art are beautiful,—as beautiful as are the lines, areas, and forms. If we can look upon the latter without relation to the real distances and shapes of physical things that they may resemble, we can just as readily survey the space of a picture without relation to the cubic yards and miles of physical space to which it may be equivalent.

And, then, what sort of an object for aesthetic contemplation would a solid sculpture or bronze be that had no resemblance to any known solid object, that, in other words, was devoid of all representative force? If it were not purely ornamental, it would tend to lack the coherence of design. If an architecture is not merely a pretty symmetrical surface, it must be welded upon the design of the building it covers, so that it expresses its functions beautifully, as a beautiful body shows the graces of the human frame. It is a distinctly inferior art that ignores the internal body and merely conceals it within a superficies of beauty. Why should not doors and windows and turrets be moulded upon a rocky hillside instead? But if they belong sensibly only to the cathedral or to the house, we must each time make an imperishable unity in beauty of the inner demands and the outer show. The designs of solid art thus inevitably find their springs in the designs our furniture, our houses, our bodies and our clothes require. Where then is the artistic irrelevance of such things of the world? How can we desire to exclude them from our art and to make its only content the beauty of ornament? It is not given to us to carve the mountains. But when they reveal their design in beautiful forms, awesome and uplifting, are they too not works of art, even though the artist be nature herself?

The only escape from this conclusion involves the acceptance of an essential irrelevance in all art, namely that what is beautiful is not what is before us in the aesthetic object, but only the fact that that object is the outcome of the artist's desire to express himself. That is out and out heteronomy. The only beautiful way in which a person can express himself is obviously to express himself in himself. Then he and his expression are one in perfect coincidence, and beautiful. But when a man makes a work of art, he makes an object that expresses itself as independently of him thereafter as his grown son ever could. In the ideal creation the artist's personality would be as completely indiscernible as is the hand of God in nature. The works of such a man would really create themselves; they would spring into being in their fundamental nucleus of purpose or design, and they would clothe themselves merely by the unfolding and complication of that first germ. We may well believe from many indications that the greatest works of art have thus come into being. The greatest artist in his greatest moments seems not to mould and to form his works but merely to yield himself to the impulses of artistic force. He is not so much a maker as a discoverer of beauty, however much he may have to grope and to search before he finds the true beauty. Its truth has no relation to the length or manner of his search. His sole task is by some means or other to find the true beauty and to recognise it then.

When the older writers said that art's function was to imitate nature, they may not have been so far from art's secret as we are now usually inclined to think. We know, of course, that the beautiful Venus is not found amongst living women and then merely portrayed. But we must not forget that she has the whole design of woman, that she expresses the ideal of woman as she lives and moves, and that her great value for us lies in that, as well as in her artistic embodiment in stone. In the Venus we do, in a sense, really 'imitate' nature; we take her design and the surface she has given it and in the whole we wrap up and preserve for our souls the elusive perfection. Would it be the same if we made an equally beautiful work of an organic type that has no counterpart or kin in nature? We not only lack the underlying design, but we do not even feel the need for its existence. When the need arises, the design soon follows and grows; and in time, with the fullness of experience and understanding, its own beauty permeates the whole.

Of course we may want to assert that we thus better nature rather than imitate her. There is no harm in this preference, provided we do not in our conceit imagine that even in our highest conceptions we have created much. We have been given almost all we have, even if we have added our labour to it. It is really absurd nowadays to keep up the foolish belief that nature is never beautiful, but only the compositions of man. On the contrary, we now feel that man is the child of nature, and that in our best moments we do but see more clearly what nature means with us in all her doings. It is not art's task, to be sure, merely to illustrate nature. But there could be no greater task for the visual arts than to discover nature's inner designs and to reveal the full harmony of these with the forms and actions into which she weaves them.

It is not given to music to reveal the ideals of our real world. But, using a Hegelian inversion, we might well claim that music's task is to reveal the realities of our ideal world. It has often been urged that music expresses much more intimately and purely the life of man's soul than any other art. How it does this may be inferred with some probability of correctness from the results of our analysis of music. We have learned that music is essentially paraphony, an interweaving of simultaneous melodic rivulets. Rhythm is a most important element in the art, though it has not been the object of our investigation at all. The motional connexions of rhythm are very powerful and combine with those of melody to form a more cogent whole than either could supply alone. The variable features of the paraphony and of the rhythm of music are, therefore, thoroughly motional. They may vary in speed and figure (rhythm in the special sense). Music may speak in one voice or in two or more. These may speak a common language, or they may utter different thoughts. Or, when the articulation of all the voices, or of all but one, is reduced in favour of their common sentiment, an atmosphere of harmony or discord in all its colours and changes may be portrayed. Loudness will indicate its strength, gentleness its peace, height of pitch its brightness or gaiety, depth its sadness or gloom.

With all these means of variation it is only necessary to bring the motions of music into some sort of correspondence with the character of the acts and energies of man for it to be able to express his soul's life. Fast or slow, vigorous or reposeful, sombre or gay, single or distressed, loquacious or reflective, clear or suspended in doubt, no mood can occur that cannot be depicted in its general character and course. And yet there is withal no representation in the art thus far.

Nor is there any symbolism or convention. We do not agree, as it were, to look upon the activities of a musical work as symbols of our soul's life. They are merely what they are,—the motions of melodies and their tonal conjunctions and changes. We enjoy them in the first place because these things are directly enjoyable by us or beautiful in themselves. But they gain in interest and passion by the fact of their natural resemblance to the 'movements' and activities of our soul as a whole, without our necessarily thereby thinking of our soul at all.

We touch here on a general problem of aesthetics that has often seemed very mysterious indeed. It is the problem of introjection, the projecting into the external phenomena of the senses of moods and sentiments that are known only in so far as they occur in ourselves. Why does a façade or a trellis look excited or calm? How can the sky look angry, or the sunset full of promise and hope? Are we not forced to believe that only some obscure analogy or association brings these notions to our minds in contemplating the object and that we then think the object looks the character that it merely makes us think of?

We may perhaps draw a better conclusion from our own case of music. Music not only seems restless upon occasion, it really is so. Now it moves at one pace, now at another. And the motional connexions of melody are as much *motions* in their own way as visual motions are in theirs. They are tonal motions : not spatial, it is true, as visual motions are, but, apart from this difference of cognitive reference, there is no great phenomenal difference. And art is concerned only with the phenomenal; it is not a practical or a scientific issue.

Again, music not only seems single or involved, it *is* so, quite as much as a train of thought or of sentiment can be single or involved. Sentiments and tones can both unite into one functional series, giving an effect of arrest or of satisfaction (of tone or of sentiment); or they may interlace in disagreement, producing a confusion that prompts us to seek their (re-)solution. And so on.

In all these respects we are merely observing and describing in the music what can also be observed and described in the thoughts, sentiments, and activities, upon which we feel we stake our whole personality. The basis of affinity between the stuff of the art in its changes and our personal mental life is therefore clearly real and indisputable.

When we feel that a person is angry with us, we do not observe that person's anger, but only his expressions. When we in turn are angry with any one, we not only feel the anger, but we express it and observe whether our expression of it seems to us to coincide with our feeling as we desire it to do. There is an inner agreement between the two that satisfies us. It has often been supposed that only at this moment do we become aware of what anger means. That is surely a mistake. It seems probable that we could equally well learn for the first time what anger is, in the expression given to it by some other person. If it were not so, a savagely angry person might well walk up and finish us off in our innocence. What less is there to note in such a first-felt anger than in the first-expressed rage of our own? Do we, this first time, merely note how our rage expresses itself, being entirely unprepared for the form it takes, and then learning what conduct signifies rage? If our expression turned out to be all smiles and compliments, should we think these were the expressions of rage? I cannot conceive that any mind could be built on so silly and witless foundations. Such a theory hands over all the dignity of coherence to the body and its heredity, making mind a pure farce. Why the body should be held to be thoroughly subject to all-pervasive law, while the mind is a mere epi-phenomenon, whose connexions even with the body are completely devoid of inner sense or of continuity with the body, I have not for long been able to appreciate.

We can therefore in all probability just as well obtain our first experience of anger from the expression (of it) that a person bestows upon us as from the acts of theirs that rouse us to an anger that we then proceed to express. We might, indeed, even encounter some emotion or sentiment for the first time in a work of art, just because that work embodied the sufficient basis of such an emotion.

From this point of view we can see not only the affinity of music with our soul's life, but also with other arts. The connexion with dancing and all its motions is obviously so close as to be almost a continuity, a complete fusion. Where any other visual display, e.g. that of the stage, can keep properly in touch with the speed of change of music, there can also be an intimate union of the two. But in opera great concessions have often to be made to the stage, so that a charge of incoherence may here be well founded. Music and sculpture or painting do not cohere at all; for the one is an art of succession and of motion, while the others are arts of simultaneity and of motionless form; and between these all true correlation of detail is lacking.

WORKS CITED

(1) ARISTOXENUS. *The Harmonics.* Ed. with trans., etc., by H. S. Macran. Oxford, 1902.

(2) BECKER, E. Die Bedeutung von Dur und Moll für den musikalischen Ausdruck. *Ztsch. f. Aesthetik u. allgem. Kunstwiss.* 1910, **5**, 216–264.

(3) BOETHIUS. *De Musica.*

(4) BRIDGE, F. and SAWYER, F. J. *A Course of Harmony.* London, 1899.

(5) BROWNE, W. D. Modern Harmonic Tendencies. *Proc. Mus. Assoc.* 1914, **40**, 139–156.

(6) COUSSEMAKER, E. DE. *Scriptorum de musica medii aevi nova series.* Vol. I. Paris, 1864.

(7) CROSS, C. R. and GOODWIN, H. M. Some Considerations regarding Helmholtz's Theory of Consonance. *Proc. Amer. Acad. Sci. & Arts*, 1893, **27**, 1–12.

(8) CURWEN, J. *A Tract on Musical Statics* [being Part A of *The Commonplaces of Music: a Student's Handbook*]. London, 1874.

(9) D'ALEMBERT, J. *Elémens de musique théorique et pratique, suivant les principes de M. Rameau, éclaircis, développés, et simplifiés.* Paris, 1752.

(10) DAY, A. *A Treatise on Harmony.* London, 1845.

(11) DESCARTES, R. *Compendium musicae.*

(12) EULER, L. *Tentamen novae theoriae musicae ex certissimis harmoniae principiis.* Petropoli, 1739.

(13) GARDINER, W. *The Music of Nature.* London, 1832.

(14) GEVAERT, F. A. *Histoire et théorie de la musique de l'antiquité.* Vol. I. Gand, 1875.

(15) —— *La mélopée antique dans le chant de l'église latine.* Gand, 1895.

(16) GEVAERT, F. A. et VOLLGRAFF, J. C. *Les problèmes musicaux d'Aristote.* Gand, 1903.

(17) GLADSTONE, F. E. Consecutive Fifths (and following 'Discussion'). *Proc. Mus. Assoc.* 1882, **8**, 99–121.

(18) GLYN, M. H. *Analysis of the Evolution of Musical Form.* London, 1909.

(19) HEFFERNAN, J. Musical Beats and their Relation to Consonance and Dissonance. *Proc. Mus. Assoc.* 1887, **14**, 1–21.

(20) HELMHOLTZ, H. L. F. V. *Sensations of Tone.* Trans. by A. J. Ellis. London, 1895.

(21) HOLDEN, J. *An essay towards a rational system of music.* Glasgow, 1770.

(22) HULL, A. E. *A Great Russian Tone Poet—Scriabin.* London, 1916.

(23) —— *Modern Harmony, its Explanation and Application.* London, 1916.

(24) —— Scriabin's Scientific Derivation of Harmony *versus* Empirical Methods. *Proc. Mus. Assoc.* 1916, **43**, 17–28.

(25) HULLAH, J. *The History of Modern Music.* London, 1862.

(26) JADASSOHN, S. *Elementary Principles of Harmony.* Leipzig, 1895.

(27) JAMES, W. *The Principles of Psychology.* Vol. I. London, 1891.

(28) KAESTNER, G. Untersuchungen über den Gefühlseindruck unanalysierter Zweiklänge. *Psychol. Stud.* 1909, **4**, 473–504.

(29) KEMP, W. Methodisches und Experimentelles zur Lehre von der Tonverschmel-
    zung. *Arch. ges. Psychol.* 1913, **29**, 139–257.
(30) KITSON, C. H. *The Evolution of Harmony.* Oxford, 1914.
(31) KRUEGER, F. Die Theorie der Konsonanz. Pt. III. *Psycholog. Stud.* 1908,
    4, 201–282.
(32) —— *Idem.* Pt. IV. *Ibid.* 1910, **5**, 294–409.
(33) LEE, V. The Riddle of Music. *Quarterly Review*, 1906, **204**, 207–227.
(34) LIPPS, T. *Psychologische Studien: Das Wesen der musikalischen Harmonie und
    Disharmonie.* Heidelberg, 1885.
(35) MACFARREN, G. A. *Six Lectures on Harmony.* London, 1867.
(36) MALMBERG, C. F. The Perception of Consonance and Dissonance. *Psychol.
    Monogr.* 1918, **25** (2), 93–133.
(37) MALTZEW, C. v. Das Erkennen sukzessiv gegebener musikalischer Intervalle
    in den äusseren Tonregionen. *Ztsch. Psychol.* 1913, **64**, 161–257.
(38) MANSFIELD, O. A. *The Student's Harmony.* 9th ed. London, 1910.
(39) MIES, P. Ueber die Tonmalerei. *Ztsch. Aesthetik*, 1912, **7**, 397–450, 578–618.
(40) MILLER, D. C. *The Science of Musical Sounds.* New York, 1916.
(41) MYERS, C. S. The Ethnological Study of Music. Among *Anthropological Essays
    presented to E. B. Tylor.* Oxford, 1907, 237–253.
(42) NIECKS, F. *Programme Music in the Last Four Centuries.* London, 1906.
(43) OETTINGEN, A. v. *Harmoniesystem in dualer Entwickelung. Studien zur Theorie
    der Musik.* Leipzig, 1866.
(44) OUSELEY, F. A. G. On the Early Italian and Spanish Treatises on Counterpoint
    and Harmony. *Proc. Mus. Assoc.* 1879, **5**, 76–99.
(45) PARRY, C. H. H. Article 'Consecutives,' in Grove's *Dict. Music and Musicians.*
    2nd ed. 1904.
(46) —— Art. 'Harmony,' *ibid.* 1906.
(47) —— Art. 'Resolution,' *ibid.* 1908.
(48) —— *Style in Musical Art.* London, 1911.
(49) PEARSALL, R. L. DE. *An Essay on Consecutive Fifths and Octaves in Counterpoint.*
    London, 1876.
(50) POLE, W. *The Philosophy of Music.* London, 1879.
(51) POTTER, A. G. *Modern Chords explained.* London, 1910.
(52) PROUT, E. *Harmony: its Theory and Practice.* London, 1st ed. 1889; 2nd ed.
    1901.
(53) RAMEAU, J. P. *Génération harmonique ou traité de musique théorique et pratique.*
    Paris, 1737.
(54) —— *Extrait d'une réponse de M. Rameau à M. Euler sur l'identité des octaves....*
    Paris, 1753.
(55) ROCKSTRO, W. S. Article 'Hidden Fifths and Octaves,' in Grove's *Dict. Music
    and Musicians* (both editions), 1878.
(56) ROUSSEAU, J. J. *Dictionaire de la musique*, 1764.
(57) SACCHI, G. *Delle quinte successive nel contrappunto e delle regole degli accompagna-
    menti.* Milano, 1780.
(58) SHINN, F. G. *A Method of Teaching Harmony based upon Ear Training.* Pt. I.
    Diatonic Harmony. Pt. II. Chromatic Harmony. London, 1904–5.
(59) —— 2nd Article 'Hidden Fifths and Octaves,' in Grove's *Dict.* 2nd ed. 1906.

(60) SHIRLAW, M. *The Theory of Harmony*. London, 1917.

(61) STAINER, J. *A Treatise on Harmony*. London, N.D.

(62) STANFORD, C. V. *Musical Composition. A Short Treatise for Students*. London 1912.

(63) STRATTON, G. M. *Experimental Psychology and its Bearing upon Culture*. New York, 1903.

(64) STUMPF, C. *Tonpsychologie*. Vol. II. Leipzig, 1890.

(65) —— Die pseudo-Aristotelischen Probleme über die Musik. *Abhandl. Ak. Wiss. Berlin. Philos. Kl.* 1896, 1–85.

(66) —— Geschichte des Konsonanzbegriffs. *Abhandl. 1. Kl. Ak. Wiss. München,* 1897, 21, 1–78.

(67) —— Konsonanz und Dissonanz. *Beitr. Akust. Musikwiss.* 1898, **1**, 1–108.

(68) —— Differenztöne und Konsonanz. *Ztsch. Psychol.* 1905, **39**, 269–283.

(69) —— *Idem*. 2ter Art. *Ibid.* 1911, **59**, 161–175.

(70) —— Beobachtungen über Kombinationstöne. *Ibid.* 1910, **50**, 1–142.

(71) —— Konsonanz und Konkordanz. *Ibid.* 1911, **58**, 321–355.

(72) —— Die Attribute der Gesichtsempfindungen. *Abh. Preuss. Ak. Wiss.* 1917 (8).

(73) STUMPF, C. und MEYER, M. Maasbestimmungen über die Reinheit consonanter Intervalle. *Ibid.* 1898, **18**, 321–404.

(74) TOVEY, D. F. Art. 'Harmony.' *Encycl. Britan.* 11th ed.

(75) —— Art. 'Melody,' *ibid.*

(76) TSCHAIKOWSKY, P. *Guide to the Practical Study of Harmony*. Leipzig, 1900.

(77) WATT, H. J. *The Psychology of Sound*. Cambridge, 1917.

(78) WERTHEIMER, M. Experimentelle Studien über das Sehen von Bewegungen. *Ztsch. Psychol.* 1912, **61**, 161–265.

(79) WESTERBY, H. The Dual Theory in Harmony. *Proc. Mus. Assoc.* 1902, **29**, 21–72.

(80) WESTPHAL, R. *Aristoxenus von Tarent*. Leipzig, 1883.

(81) WOOLDRIDGE, H. E. *The Oxford History of Music*. Vol. I: The Polyphonic Period. Oxford, 1901.

(82) ZARLINO, G. *L'istitutioni harmoniche*.

(83) Patrologiae cursus completus. Series Latina prior. Ed. J. P. Migne. Paris.

INDEX OF AUTHORS

D'Alembert, J., x, 36 f., 66, 99 f.
Aristoteles, quisdam, 186
Aristotle, 6, 38, 51, 82 f., 224
Aristoxenus, xi, 7, 107, 158

Bacchius, 157
Becker, E., 167
Boethius, 38
Bridge, F. and Sawyer, F. J., 179 f.
Browne, W. D., 166

Cherubini, M. L., 92
Cotto, J., 82
Cross, C. R. and Goodwin, H. M., 29
Cummings, W. H., 86
Curwen, J., 39

Day, A., 42, 99, 116 ff.
Debussy, C. A., 16, 140
Descartes, R., 23, 112
Dunstable, John of, 86

Elert, K., 120
Euclid, 38
Euler, L., 38

Gardiner, W., 12
Gaudentius, xi, 15, 108, 157 ff., 213
Gevaert, F. A., 38, 51, 52, 84 f., 108, 146, 155, 157, 159
—— and Vollgraff, J. C., xi, 6 f., 52, 82, 85, 154, 157, 167, 178, 184, 224
Gladstone, F. E., 84 ff., 88, 93 ff., 117 f., 120
Glyn, M. H., 149
Guido, 85

Hauptmann, M., 142
Heffernan, J., 24, 33
Helmholtz, H. v., vi, 3, 24 ff., 32 f., 69 f., 87 ff., 146 f., 162, 176 f., 184, 186 ff., 224
Holden, J., 38, 175
Hornbostel, E. v., 153
Hucbald, 85
Hull, A. E., 16, 94, 117, 120, 140, 159, 166
Hullah, J., 63 f., 111

Jadassohn, S., 125 f., 128 f.
James, W., 69
Johannes de Garlandia, 159, 187

Kaestner, G., 196
Kemp, W., 142, 175 f., 186
Kitson, C. H., 124, 126, 128
Kollmann, A. F. C., 94
Krueger, F., 186 f., 195
Külpe, O., 175, 186

Lasus, 7, 38
Lee, V., 225
Lipps, Th., 26, 38

Macfarren, G. A., 3, 51, 85 f., 89, 93, 97 f., 116 f., 133 f., 136, 166, 175, 192
Mach, E., 100
Malmberg, C. F., 195 f.
Maltzew, C. v., 52 f.
Mansfield, O. A., 117, 135 f.
Meyer, M., 42, 188 f.
Mies, P., 226
Miller, D. C., 3, 27
Myers, C. S., 34

Nicomachus, 38, 154
Niecks, F., 226

Oettingen, A. v., 142
Ouseley, F. A. G., 83, 99

Parry, C. H. H., 104, 116 ff., 162 f.
Pater, W., 225
Pear, T. H., 175
Pearsall, R. L. de, 95, 117
Plato, 83
Plutarch, 51
Pole, W., 4, 84 ff., 89, 163
Potter, A. G., 172
Prout, E., viii f., 35, 99, 102 ff., 111, 116 f., 121 ff., 128 ff., 133 ff., 172 f., 177, 184
Ptolemy, 107
Pythagoras, 38

Rameau, J. P., ix f., 23, 36 f., 65 ff., 79, 99,
    142, 165
Riemann, H., 142
Rockstro, W. S., 122
Rousseau, J. J., 39
Russell, B., 100

Sacchi, G., 66 f., 86, 88 f., 90 ff., 118 f.
Schubert, F., 34 f.
Scott, C., 132, 140
Scriabin, A., 16, 80
Serre, J. A., 28
Shinn, F. G., 95 ff., 116 ff., 120 f., 125 ff.,
    135 f., 170 f.
Shirlaw, M., ix, 23 f., 28, 66 f., 142, 149,
    182, 184
Sorge, G. A., 28
Stainer, J., 117
Stanford, C. V., 162

Stephens, C., 85, 89
Stratton, G. M., 49 f.
Stumpf, C., xi, 6, 15, 23 ff., 32, 42, 51, 70,
    83 f., 108, 110, 142 ff., 153 f., 156 ff.,
    164, 184 ff.

Tartini, G., 23, 28, 142, 224
Tchaikovsky, P., 106, 116, 123 ff., 128 f.,
    135 f.
Thrasyllus, 157
Tovey, D. F., 54, 129, 151, 164, 182

Wagener, A., 84
Wertheimer, M., 78
Westerby, H., 142
Westphal, R., 7
Wooldridge, H. E., 51, 82 f., 92, 159

Zarlino, G., 23, 87, 89, 142

INDEX OF SUBJECTS

Abstraction, 73 f., 77, 83, 87
Aesthetics as science, vi, 98 f., 131 ff., 221 ff.
Analysis and synthesis, 1 ff., 55 ff., 143 f., 202 f.
Art and science, 22, 55, 60, 64, 101, 139 f., 150, 214 f.
Ascent and descent of pitch, 52, 127, 131, 184
Attributes of sound, 5 ff.

Balance, 8 f., 19 f., 57, 72 ff., 79, 200; of parts, 97; of distinction, 147 f.
Bass most prominent, 51, 92, 102, 130
Beats, 24 ff., 26 f., 31, 56, 180 f.
Beauty objective, 216 ff.
Blend, 5, 56 ff.
Breadth of tone, 7, 13 f.

Chords, theory of, 148 ff., 163 ff., 209 ff.
Cochlea, 46 f., 203 f.
Columns of thirds, etc., 79 f., 148 ff., 163, 209 f.
Concordance, 141 ff.
Consecutives, theories about, 81 ff.; facts about, 101 ff.; explanation of prohibition, 109 ff.; exceptions to prohibition, 115 f.; 171 f., 207
Consonance, grades of, 15 ff., 83, 94, 184 ff., 223
Consonance and dissonance, 15 ff.; Helmholtz's theory criticised, 24 ff.; Stumpf's theory criticised, 146 ff.; 151 ff., 155; in successive tones, 25, 31 ff., 33 ff.
Convention or habit in music, 84 ff., 89, 93 f., 96, 132, 216 f., 230

Development in music, 15 f., 62, 92, 96, 113, 150, 152 ff., 159, 161 ff., 168, 196 f., 212 f.
Difference tones, 28 ff., 31, 56, 58 f., 192 f., 196, 202, 224
Distinction, loss of, 147 f.

Ear, training, 179 f.; absolute, 189; verdict of, 204

Exposed intervals, v. hidden, 125 f., 136 f.
Exposure, rhythmical, 134, 137 f.
Expression in music, 229 ff.

Factors, positive and negative, in beauty, 89, 96, 127 f., 173 ff.
Field, the auditory, 48 ff., 53 f., 201 f.
Fifth, the, 19 f., 29, 33 f., 59; a bare interval, 83, 96; fifths not equivalent, 77 ff.; consecutive, 80 ff.
Form and mass in hearing, 43 f.
Fourth, the, 21, 133 ff., 177 f., 191 f., 210; as dissonance, 136 f.
Fusion, 11, 15 ff., 18 ff., 33 ff., 43, 62 f., 64, 74, 78, 94, 96, 173 f., 190; explanatory value of, 141 ff., 185 ff.; grades of, 16 f., 21 ff., 32, 71, 102 ff., 110 ff., 122 ff., 147 ff., 151 f., 156, 184 ff., 200 f.; neutral grades of, 108, 111 f., 147 f., 187, 195

Greek music and theory, 6 f., 15, 18, 51, 81 f., 92, 107 f., 110, 146, 151, 154 ff., 167, 178, 185, 206, 213

Harmonics, v. partials
Hidden intervals, 122 ff.; v. exposed
Highest tones, 12, 45 ff.
Horizontal and perpendicular, 63 f., 111 f.

Imagination, force of, 90 f.
Individual differences in ear, 61
Instrumental tones, 1 ff.
Intensity, 6, 8, 12, 19, 72, 198 f.
Interval, 17, 36 ff., 200; in inversions, 71 f., 173 ff.; cause of precision of, 192 f.; nomenclature of, 16; ascending and descending, 52 f.; simultaneous and successive, 53; intrinsic character, 95 ff.
Inversion of chords, 66 ff., 76 f., 88, 95, 149, 151 f., 175 ff., 203, 209

Longitudinal and transverse, 13 f.
Lowest tones, 12, 45 ff.

Major and minor triads, difference between, 52, 142, 166 f.

Materialistic theories, prejudice for, v, 204

Melody, 205; statistics of, 34 f.; in bass voice, 51 f.; as surface of music, 54, 61; as basis of prohibition of consecutives, 111 ff., 126; of hidden or exposed intervals, 128 ff.; of fourth from bass, 134 ff.

Melodic distinction, 148, 156, 160 ff.; flow, 172, 178, 182, 206, 208 f.; steps and leaps, 34 f., 124 f., 128 ff., 134 ff.

Memory, 27 f., 31, 42, 194, 196

Method, 101 f., 109 f., 122

Motion, contrary and similar, 82, 170 ff., 207 f.

Mystery in music, 140

Objectivity of beauty, 214 ff.

Octave, the, 19, 33, 42 f., 59; equivalence of, 65 ff., 96, 113 f.

Order or position, 7, 48 ff., 198 f.

Organum, 82 f.

Overlapping of tones, 10 ff., 18 ff.

Paraphony, 155 ff., 187 ff., 206 ff., 230 ff.; as basis of polyphonic music, 160 ff.; factors that modify, 169 ff.; interaction of, 173 ff.

Partial tones, viii, 1 ff., 15 f., 149, 166 f., 180 f., 192 ff., 196, 202; relation to fusion, 25 ff., 31; in synthesis and analysis, 56 ff., 61, 67 ff., 76 f., 87 ff.

Particles of sensation, 50; of sound, 9 ff., 18 ff.

Parts, the number of, 130 f.

Pattern, 72 ff., 145, 165, 203, 211

Pitch, 4, 7 ff.; rising as departure, 52; musical range explained, 45 ff., defined, 47; relation to volume, 8 ff.; central in volume, 9 ff., 19, 199

Pleasure in relation to beauty, 217

Polyphony and harmony, 43 f., 62 ff., 111, 162 ff., 205

Primary tones in analysis and synthesis, 60 ff.

Proportion, 36 f., 73

Pure psychology, 5, 20 ff., 37, 54, 98 ff., 221 ff.

Pure tones, 3 ff.

Quality, 5; amongst the tones of an octave, 70 f., 78

Relationship of chords, 94 ff., 117 f., 120 f., 127 ff., 184, 212

Representation in music, etc 225 ff.

Resolution, 137, 182 f., 211

Rest and motion, musical, 151 ff., 155

Rhythm, 91, 137, 178, 230

Rules in music, 81, 84 ff., 132, 216 f., v. convention

Sensations, 5

Seventh, diminished, 106

Sixth minor and augmented fifth, 180 ff., 210

Smoothness in tone, 8 f., 25, 57

Sonorous body, the, ix, 66 f., 149

Statistics, viii, 34 f., 115, 117 ff., 139 f.

Subjectivism, 214 ff.

Surface of tone, 57 f.

Symmetrical, 8 f.

Synergy, 23 f., 32 f., 70

Thirds and sixths as consonances, 107 f., 142, 147 f., 154, 157 ff.

Tonality, 53, 90 ff., 164 ff., 183 f., 211 f.

Tones and noises, 8

τοπος, 7

Trend of opinion, 104, 115, 123, 128, 131 f.

Triad, theory of, 148 ff.

Tritone, 103, 105 f.

Unconscious, the, 59

Unison, the, 130, 151

Vision, compared with hearing, 33, 40 f., 43, 48 ff., 73, 85

Voices, relative prominence of, 51, 92, 102 ff., 122 ff., 128 f., 208; effect of number of, 127 f., 169 f., 208; 110 f.

Volume, 6 ff.; inversely proportional to ratio of vibration, 20; graded in intensity, 72, 198 f.

Western music, early, 51, 82 f., 151

Zones, 93 f.; of paraphony, etc., 188 ff.

For EU product safety concerns, contact us at Calle de José Abascal, 56–1°,
28003 Madrid, Spain or eugpsr@cambridge.org.

www.ingramcontent.com/pod-product-compliance
Ingram Content Group UK Ltd.
Pitfield, Milton Keynes, MK11 3LW, UK
UKHW051009240426
470322UK00018B/571